Adam's Nose,
and the Making of
Humankind

Adam's Nose,
and the Making of
Humankind

Michael Stoddart

University of Tasmania, Australia

Imperial College Press

Published by

Imperial College Press
57 Shelton Street
Covent Garden
London WC2H 9HE

Distributed by

World Scientific Publishing Co. Pte. Ltd.
5 Toh Tuck Link, Singapore 596224
USA office: 27 Warren Street, Suite 401-402, Hackensack, NJ 07601
UK office: 57 Shelton Street, Covent Garden, London WC2H 9HE

British Library Cataloguing-in-Publication Data
A catalogue record for this book is available from the British Library.

ADAM'S NOSE, AND THE MAKING OF HUMANKIND

ISBN 978-1-78326-517-6
ISBN 978-1-78326-518-3 (pbk)

Typeset by Stallion Press
Email: enquiries@stallionpress.com

Printed in Singapore

About the Author

 Michael Stoddart's interest in the sense of smell started when he was conducting his PhD study on the population dynamics of the European water vole. Large paired scent glands lying on the animals' flanks waxed and waned with the rodents' reproductive cycles and when bad weather precluded fieldwork, Michael turned his attention to the structure, function and physiological control of the glands. This early interest laid the foundation for much of his future research. Over several decades he has examined how mammals use their sense of smell, how smells control social behaviour, how they modify the body's hormone levels and influence reproductive physiology.

Adam's Nose, and the Making of Humankind is Michael's fourth book on mammalian olfaction and the second on the human sense of smell. It draws on groundwork laid in his textbook *The Scented Ape: the Biology and Culture of Human Odour* (CUP 1990), and puts recent olfactory genetics research into the context of human evolution.

Michael is a Scot, born in Lanark and brought up near London. After leaving school he studied zoology at the University of Aberdeen where he gained BSc and PhD degrees, and later a DSc. After a post-doctoral position at the Animal Ecology Research Group in the University of Oxford he was appointed Lecturer, later Reader, in Zoology at King's College London. In 1985 he was appointed Professor of Zoology at the University of Tasmania in Hobart, Australia. Using the social marsupial sugar glider as a model species he studied the endocrinological basis of scent production, and how the gliders' scent influences stress and social behaviour. He spent five years as Deputy Vice-Chancellor at the University of New England in New South Wales, and ten years as Chief Scientist of Australia's Antarctic program, before returning to the University of Tasmania to establish the University's Institute for Marine and Antarctic Studies. He is currently Professor Emeritus in the Institute.

The Soul of the Rose John William Waterhouse 1908 Bridgeman Art Library

Contents

Introduction

'*Adam's Nose and the Making of Humankind*', including Eve's nose, of course, is the story of how the sense of smell evolved from the earliest beginnings of life on Earth to grace the only animal on Earth to have developed a smell culture. Along the way it was subjected to a genetic mutation that crippled part of it, completely and irrevocably changing the lifestyles of our distant ancestors. That mutation let ancient primates climb down from the trees to go out onto the brightly lit African grasslands, away from an old life spent searching for berries, fruits, nuts, and occasional frogs and lizards, to a new one offering immense new dietary riches. The herbivorous mammals that lived on the plains were too large and too fast for a single individual to hunt. The only way they could be exploited was if several individuals worked together to hunt collaboratively. To ensure that there was always sufficient numbers of hunters available, the new plains dwellers lived communally, forsaking long held practices of territorial ownership. Communal living also allowed the spoils of the hunt to be shared with others in the community, sowing the seeds for what eventually would develop into social dining. Crucially, the mutation let them live as they always had done, in families we would recognize today as similar to our own. Living communally so they could enjoy the fruits of their collaborative labour didn't mean they had to live in an amorphous herd; the family unit remained paramount to the way they organized their lives. That little mutation turned the ancient ape into a hunter and a carnivore, turned an ape into a human and fashioned a species that was to go on to dominate the world.

This book is about the context in which the human sense of smell evolved, about how it works and how animals use their noses in their daily lives. Within this context, the things for which we use our noses seem remote from the important issues of survival. Every species of animal on Earth uses smell for just about everything it does in life, from choosing a place to live, finding food, detecting predators, and for finding a mate — the ultimate imperative without which the existence of their line would end. Every species, that is, except *Homo sapiens*. Our sense of smell seems to serve sensual self-indulgence for the pursuit of pleasure; it isn't much given to satisfying the necessities of life.

We modern humans have a complicated relationship with our noses for which we have to thank our animal origins. We don't want to smell human, yet we spend huge sums of money so we can subtly smell of animals. In our ultra-scrubbed and scentless world we Westerners experience life as though it's divorced from nature, and not without reason. Our food's all too often processed and packaged with little hint of the animals and plants it comes from.

Pharmaceuticals control an ever-increasing number of diseases, letting us live in ways undreamed of by our ancestors. Clothes are spun from synthetic fibres owing nothing to cotton or wool, and our dwellings are increasingly built from man-made materials. The view that nature is for the animals, and that humankind has shaken off its animal origin is reinforced everywhere we look. This book will show we have shaken off no part of our animal origin and that we're much more like animals than we might be comfortable acknowledging; where our raw animal blood is exposed is in our strangely hobbled sense of smell.

*   *   *

We can't help but view the world through our five main senses — sight, touch, hearing, taste, and smell. We've no other way to do it. Most of us would say our senses are dominated by our eyes and ears, and to an extent that's correct. Most people regard their noses as being rather poor; you wouldn't want to rely on your nose for deciding when it's safe to cross the road, for instance, but for enjoying a bowl of fragrant rice, or for gladdening your heart when a shower of rain has breathed life into a summer garden, *only* your nose will do. Smell is the true sense of indulgence, whence springs the notion of 'sensual'.

Scents and smells appeal to the hedonist in each of us; scents that seduce, excite and tease, smells that uplift, empower and fortify, and odours that challenge, test and taunt. Yet, in denying that we're animals smelling of humans, are Adam's and Eve's noses merely the desiccated remnants of a once all-powerful system from which most of the juice has been sucked, like a small country clinging to the tattered shreds of Empire while heading towards extinction, or are they well-adapted organs, enabling their bearers to better survive the present, and to successfully tackle the future? This book will show they are far from the exhausted remnants of former glory and that they live in our lives in unexpected ways.

I've worked in the field of smell biology for over four decades, on and off, and can more than hold my own at every cocktail and dinner party to which I'm invited, because everyone's fascinated with smells and scents, and eager to know more. They want to tell me their own smell stories, often with an innocence suggesting they have a view their sense of smell is so personal its sensations can't conceivably have been experienced by anyone else. Unless I've been to the wrong dinner parties, few people seem as interested in visual and hearing phenomena as they are in what their noses tell them. It appears we take playing football, or typing a manuscript as I'm doing now, much more for granted than we do the beguiling scent of a summer rose, the magical aroma of a fine Barossa Shiraz, or the smells of Grandmother's kitchen, redolent with memories of yesteryear.

Innocent and comforting as these things are, sex is never far away from smell. Annick Le Guérer, the distinguished French scholar who has written the most comprehensive, and most readable review of how philosophies about smell from Greco-Latin times have changed, describes Plato as making the distinction between 'those senses capable of providing pleasures that elevate the soul' (the eyes and ears), and 'those that are the source of purely carnal indulgence' (the nose).[1] In Plato's day, anything that led humankind towards concupiscence was frowned upon, like animal smells and prostitutes. The macho Ancient Greeks vilified scents and perfumes, branding their users as effeminate. They deemed perfumes as fit only for prostitutes, though strangely they didn't pause to question *why* prostitutes used them, nor *how* perfumes helped them drum up custom. The Romans took a more pragmatic view. Various Caesars tried to stop people using perfumes, not because they thought them effeminate, or that imperial collapse might follow, but to try to protect the state from runs on the public treasury — fine perfume has always been expensive. As it happened, the huge outflow of money from Roman coffers to pay for spiralling fragrant decadence was a contributory factor in the eventual decline of their mighty Empire. In its heyday, the use of perfumes and fragrances by the Romans eclipsed anything seen before or after — at least until the decadence of the present time.

The early Christian church endorsed the Greek philosophy of separating the body from the mind, being quick to denigrate anything that ornamented the body to the detriment of the mind. The fine scents of incense were for God's pleasure; diverting them from the pious use for which they were intended ran counter to the struggle against concupiscence. Only through chastity could the flesh be made incorruptible, and only through incorruptible flesh could scents and smells become spiritual. Throughout recorded history, and across all cultures, the association between smell and spirituality has remained strong and vibrant, though it appears chastity has had little to do with it. The line between the sacred and the profane blurs easily it seems, when smells are involved.

The Age of Enlightenment saw a new interest in the sense of smell as philosophers started to study the human body. A distinction between aesthetic and unaesthetic senses was made in an attempt to separate the good from the bad; Kant and Hegel as late as the 18th century maintained that smells were antisocial, and therefore unaesthetic. Kant went so far as to say that smell was the most dispensable sense, because

---

[1] Le Guérer, A. 1994 *Scent. The Mysterious and Essential Powers of Smell.* Chatto and Windus, London.

disgusting smells vastly outnumbered pleasant ones. He must have had a poor sense of smell, or perhaps he lived too close to an abattoir! There was Enlightenment support, too, for Plato's earlier position on perfumes and prostitutes. Smell was regarded as an animal sense driving what generations of biology students learn as the four essentials of animal behaviour — feeding, fighting, fleeing, and copulating — above which humankind must rise to become fully human. Yet from time immemorial the allure of scent, particularly in sexual attraction, has been deeply engrained in the human psyche. Attitudes towards smell in the 19th century's societies of Europe and America were *appliquéd* onto the sins of humankind's evolutionary origins, creating tension where none should exist. In the East, however, the nose was celebrated with style and flourish, without attempts being made to demonize what human bodies smell like.

Constance Classen and her colleagues have examined the cultural history of smell most thoroughly in their book: *Aroma. The Cultural History of Smell*.[2] In the 18th and 19th centuries Western scientists and intellectuals devalued and feared smell, promoting instead vision as the culturally supreme sense. Smells were given cultural values that held the various classes of society in their allocated places. As Classen puts it:

'This powerful denigration of smell by Europe's intellectual elite has had a lasting effect on the status of olfaction. Smell has been "silenced" in modernity. Even on those rare occasions when it *is* the subject of popular discourse... it tends to be presented in terms of its stereotypical association with moral and mental degeneracy....

'Why this cultural repression and denigration of smell? Generally speaking, those elements which are systematically suppressed by a culture are so regulated not only because they are considered inferior, but also because they are considered threatening to social order.'

It took until the middle of the 20th century for the 'olfactory silence' to be broken and today there is a new enlightenment about smell. If you search on the internet for the words 'sight', 'hearing' and

---

[2] Classen, C., Howes, D., and Synnott A. 1994 *Aroma. The Cultural History of Smell*. Routledge, London.

'smell' you will be greeted with 150 million references to the first, 115 million to the second and 87 million for 'smell'. There's a bit of catching up yet to be done, but the silence is broken.

The repressive social attitudes of 17th, 18th, and 19th century Europe towards smells sit oddly alongside some of the world's most arresting poetry, literature and classical art. The English poet Robert Herrick wrote evocatively of the scent of his various lovers, of which he had quite a few, while the Frenchman Joris-Karl Huysmans shocked genteel Parisians with descriptions of the scents of the armpits of hard-working country women, and of the genteel women in the grand ballrooms of Paris. Perhaps the most beautiful of all, the *Song of Solomon*, is the supreme celebration of the perfume of a beloved, in which the writer compares the scent of his betrothed to the most beautiful things on Earth. Many 19th century painters even captured the scent of their subjects in brush strokes and pigment — John Roddam Spencer Stanhope's painting *Eve Tempted* is an excellent example, among many in a similar genre. Western societies have art galleries and concert halls for the social celebration of visual and aural arts, and libraries for literature, but there is no place for mass odour culture — unless you count the great cathedrals as theatres to bursts of billowing incense. In only a few cultures is there any semblance of a popular odour culture; the Japanese incense ceremonies flourished for centuries until Westerners derided them as effeminate and not the thing for 'real' men; such ceremonies are nowadays maintained largely for tourists and are no longer part of mainstream Japanese life. It seems the Western smell-malaise, firmly grounded in ancient Greek philosophy, eventually infected the East.

In many non-European cultures, smell is central to a person's identity and is reflected in greetings ceremonies. Among the Gidjingali of Australia's Arnhem Land, for instance, the traditional greeting practice was for the host to take sweat from his own armpit and rub his hand, first under the armpit of his guest and then back to his own; the ritual of greeting involved the sharing and mixing of personal essences.[3] Mutual rubbing of noses, while imbibing deeply of the

[3] Eibl-Eibesfeldt I. 1977 *Society and the Modern World*. Australian National University, Canberra, Australia.

other's scent, is still commonly practiced among certain Pacific Islanders and for the Maori of New Zealand is a standard form of formal greeting. The hypnotic scent of frangipani blossoms, made up into garlands, greet visitors to many remote Pacific communities, enveloping the giver and recipient in one fragrant cloud. By comparison, the remote handshake of Westerners is olfactorily *sense*-less — what a pity the elegant kiss to the back of a perfumed hand has fallen from fashion!

In contemporary literature there is an emerging genre that deals with how people interact with smells, both pleasant and unpleasant. The long-established view that women are more attracted to floral fragrances than to musky animal scents belongs to the 18th and 19th centuries, and helped define femininity. The renowned French sociologist, Alain Corbin, whose 1986 book *The Foul and the Fragrant* has gone far to debunk the prevailing views of the European intelligentsia about the place of smells in society, draws attention to the traditional relationship between flowers and femininity:

> '... the natural, sweetly perfumed woman-flower disclosed a firm wish to control emotions. Delicate scents set the seal on the image of a diaphanous body that, it was hoped, simply reflected the soul... She should be rose or violet or lily — certainly not feline or musky; floral images supplanted those borrowed from the carnivorous cycle.'[4]

Inspired by Corbin's writings, contemporary authors are challenging the traditional stereotype. Danuta Fjellestad, of the University of Uppsala, has analysed smell symbolism in a number of contemporary novels to reveal how smells 'borrowed from the carnivorous cycle' are increasingly being used to define the individuality of female characters. Quoting from Jeanette Winterson's novel *Written on the Body* (1993), she notes the main female character's olfactory universe is 'a mixture of excremental and incense odours; she is "a dark compound of sweet cattle straw and Madonna of the Incense. She is

---

[4]Corbin, A. 1986 *The Foul and the Fragrant: The Sense of Smell and its Social Image in Modern France*. Picador, London.

frankincense and myrrh, bitter cousin smells of death and faith'".[5] Fjellestad illustrates how the use of olfactory imagery, and exploration of the role of smell in sexual attraction, is challenging long-standing conventions. It's only in our species — the only one that doesn't use smells for just about everything in life — that culture has tried to determine and dictate how people should smell, and what they should think about the smells of others. To the animals of the world, this is craziness beyond belief.

* * *

Humankind is the product of a long evolutionary process, even though we've been distinguishable from other primates for less than half a million years — hardly a blink in the 3.4 billion year history of life on Earth. As a recognizable entity, our species came but lately to Earth's fauna. To give a perspective on just how lately, imagine the height of New York's Empire State Building, from the street to the observation deck on the 102nd floor 1,224 feet up, as representing the 4.6 billion year time span from when Earth was formed from galactic dust and rubble. Imagine time travellers climbing up the staircase, looking in at each floor (Figure I.1). The earliest bacteria appear on about the 27th floor, some 320 feet above the street. For the next 63 floors nothing much new seems to happen; life just ticked along in its comfortable primeval ooze. Then, as the travellers reach the 90th floor, 1,080 feet above the street and with 88% of their journey behind them, things start to hot up. Around the time periods known as the Ediacaran and Cambrian, between about 580 and 485 million years ago, there was an explosion of life, and for the next 12 floors, animal groups and species come and go, some flamboyantly, others quietly. Our own species' sojourn has lasted less than one twenty millionth of the whole history of Earth. That means we made our entry into the Empire State Building at the level of the observation deck, and we have been on Earth for a length of time

---

[5] Fjellestad, D. 2002 Towards and aesthetic of smell, or, the foul and the frangrant in contemporary literature. CAUSE: Revista de Filiogía y su Didáctica 24: 637–651.

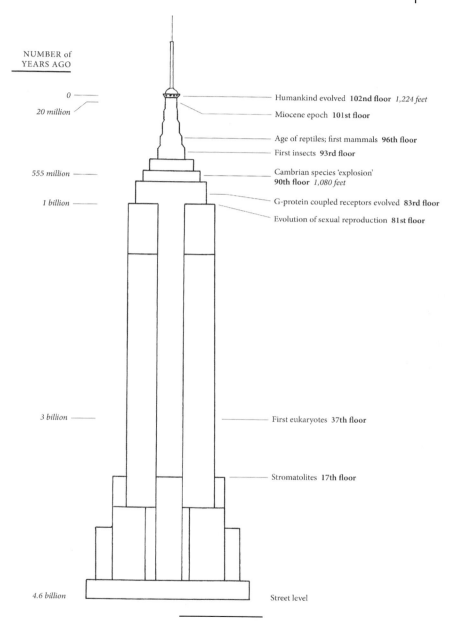

NUMBER of YEARS AGO

0 ——— Humankind evolved **102nd floor** *1,224 feet*
20 million ——— Miocene epoch **101st floor**

Age of reptiles; first mammals **96th floor**
First insects **93rd floor**

555 million ——— Cambrian species 'explosion'
90th floor *1,080 feet*

1 billion ——— G-protein coupled receptors evolved **83rd floor**

Evolution of sexual reproduction **81st floor**

3 billion ——— First eukaryotes **37th floor**

Stromatolites **17th floor**

4.6 billion ——— Street level

Figure I.1

Sketch of the Empire State Building, New York, annotated to show significant milestones in evolution.

equal to the thickness of the pile of the carpet under our feet! Everything humankind has created and destroyed — all the love, all the music, literature, architecture, art, all the scientific and medical advances, all the hatred, all the suffering, all the wars — reduced to just the thickness of the carpet pile on the 102nd floor of the Empire State Building! It is on the 101st and 102nd floor that the sense of smell made its leap from a fundamental part of life that enabled our lineage to give us breath, to something intriguing, enigmatic and unique in the animal kingdom.

*   *   *

I'm a biologist trained in classical evolutionary zoology. Charles Darwin's theory of evolution through the natural selection of characters best suited to a changing environment excites me now as much as it did when I was a student, despite my having read widely about Creationism, Grand Designers, Natural Theology and faith. Its simplicity and logic are compelling and stunning. The palaeontological evidence supporting it is strong. The flowering of molecular genetics in the last few decades has enabled the forging of some of the evolutionary links Darwin intuited, but had no way of understanding. The unravelling of a species' genetic heritage allows us to look back in evolution, to before the time the species first appeared in the fossil record, letting us read the genes of the last common ancestor it shared with another species. The full power of the tool is yet to be revealed and when it is, many of the evolutionary conundrums with which science has been wrestling for a century and a half will be resolved.

Despite my support of Darwin's theory of evolution by natural selection, I don't support the view that human biology can be interpreted through the study of animals alone. As Jon Mooallem put it in the New York Times magazine: 'It's naïve to slap conclusions about a given species onto humans. But it's disingenuous to ignore the possibility of any connection.'[6] Wise words indeed. If you accept Darwin's theory of evolution through natural selection as the best

---

[6]Mooallem, J. 2010 Can animals be gay? New York Times Magazine 31st March.

model to explain our origins, you have to accept the existence of a process, or processes, that chooses the genes to be road-tested in the next generation. If you accept that humans are the product of Darwinian evolution, study of what went before we came along provides the building blocks for understanding our origins, and for what may come after. If you don't accept that humankind is the product of Darwinian evolution, you'll likely find much to rile against in the following pages.

*　　*　　*

Adam's Nose and the Making of Humankind starts with a statement about our enigmatic relationship with smell; we've a primeval urge to smell of something, but not to smell of the humans we are. We evolved from species living in mostly monogamous small families, yet our history shows we've lived communally for most of our existence, and now we live in vast conurbations of many millions of souls. Of the greatest significance to our story is the fact that humankind is the *only* one among the 5,400 mammal species to live communally while maintaining a mostly monogamous lifestyle, and we'll investigate how this comes about.

Before examining how noses, snouts and the vast array of smell receptors found in the animal kingdom work, some pages are given to describing how smell works, setting out the chain of events occurring between a smell molecule being swept up into your nose and a signal reaching your brain. The smelling brain is described in enough detail to provide a basis for an account of the close links between smell and sex, and how sex is coordinated in animals. The parallels between how sex happens in sea squirts (animals you may not be very familiar with) and us are striking. Sea squirts occupy a pivotal point in the evolutionary story, and study of them reveals how breathing air has separated the nose from the part of the brain where sex is coordinated. The steps in the smell–sex link in the sea squirt are fundamentally the same as in ourselves. Comparative anatomy, as this sort of investigative approach is called, has much to offer.

Smell governs sex and sexual behaviour in mammals, showing its widespread Svengali-like control over the animal kingdom's most basic imperative. The singular difference between humans and mice, at least in the context of sex, is that human sexual biology is profoundly influenced by the genetic mutations that affect parts of our sense of smell. Human body odour is generally not considered particularly attractive, at least during casual acquaintance, as any commuter packed into a hot, overcrowded train will aver. This is surely truest for the smell of armpits. Armpits are the seats of olfactory information about a person's genetic makeup, though there's no clear evidence that modern humankind makes any use of that information in the most important biological task of all, namely in the selection of our mates. Why armpits are routinely attacked with soap, water and multi-bladed razors, before being smothered in perfumed 'deodorant', doesn't tell us that armpits are biologically redundant. It reminds us that humankind is evolutionarily very young and its model is still under construction.

Incense and perfume come next; their universal appeal rests on the smells of animal products and plant exudates that remind us of animals. Sensuality is the hallmark of fine perfumes. Artists, writers and advertisers make subtle use of smell imagery in their works, and through their skills deflect the enquiring mind away from the real underlying message — one that's more animal than aesthetic. Finally, the various threads running through the journey are drawn together in an exploration of how our sense of smell — and what's happened to it along the way — made us human.

I'll show evolution has removed from humankind many instinctive reactions to smells to which other animals remain slaves, inviting our forebears to replace those smells with a rich and pervasive smell culture. Removal of instinctive reactions made it possible for us to live in close proximity to so many of our fellows in relative, if not total, harmony. Gratification of the sense of smell percolates through everything we do, from the planting of fragrant gardens, to the enjoyment of fine wines; from the use of rich perfumes, to the gentle arts of seduction and *haute cuisine*. Sight, hearing, touch, and taste have much routine and mundane work to do; to smell falls the shimmering mantle of sensual pleasure.

Because you've read this far I'm assuming you're a person who's fascinated by what it means to be a human being. I'm also hoping you're someone who's interested in science and wouldn't be uncomfortable reading a weekly mass-circulation science magazine, because the story of smell is largely a scientific one. Our journey towards understanding how Adam's and Eve's noses have made us human will take us through some scientific briar patches. I'll cover these as simply and clearly as I can, and suggest you stay close as we negotiate some of the trickier bits. I've presented only the indispensable minimum of information necessary for coherence such that a PhD isn't a mandatory requirement for following the journey! I haven't written *Adam's Nose* in the *staccato* form of a scientific textbook, interwoven with cross references to the published literature; rather I've provided a few key suggestions, in footnote form, for those who want to go off on their own side trips. I hope you'll feel encouraged to explore for yourself.

Chapter *1*

The Enigma of Smell

On a balmy summer evening, a young noblewoman prepared herself for a first date with a warlord, a powerful soldier-politician in charge of a huge tract of land, and a ruler over great multitudes. He needed an alliance with her for the army she controlled; she needed him for the influence he could exert to protect her country from external interference. She planned her meeting meticulously. Although not beautiful in a classical sense, she knew exactly what powers of persuasion were needed to make this a meeting of equals. She was experienced in these things, having a few years earlier seduced the most powerful man in the world. So she bathed in milk and honey, applied fragrant oil and the most precious perfumes to her skin. Servants painted her feet and hands with camphor so every step she took, and everything she touched, lingered of her scent. She sucked pastilles of fragrant resin to sweeten her breath and painted her eyes with the darkest kohl, redolent with the brooding fragrance of sandalwood. She would make her appearance by barge, gilded and burnished to reflect the rays of the setting sun. Young boys dressed as Cupids would spray its purple sails with rosewater; the deck would be strewn with flowers and fragrant herbs. Braziers on the bow consumed frankincense, myrrh and cassia, delicate scents that wafted the far-away shore with expectation. The soldier-politician, a man with a proclivity for feminine charms, was drunk on her aroma long before his eyes met hers. The quay quivered with excitement as the barge hove to.

1

And so Cleopatra met Marc Antony, in one of the most famous encounters in all of history.[7]

Cleopatra gauged that her powers of persuasion, in addition to her looks, intellect and wit, must include a personal odour statement — a scent signature — every bit as much hers as the dimples in her cheeks and the swing of her hips. For his part Marc Antony, too, undertook a lengthy toilet. The Greek scholar Antiphanes records the preparations he might have undertaken:

'He really bathes
In a large gilded tub, and steeps his feet
And legs in rich Egyptian unguents;
His jaws and breast he rubs with thick palm oil,
And both his arms with extract of sweet mint;
His eyebrows and his hair with marjoram
His knees and neck with essence of ground thyme.'[8]

Powdered orris, storax, labdanum, marjoram and nard were sprinkled among his linen before he dressed, so he exuded a subtle, manly scent — not sweet, but strong and robust. At his first banquet with the young queen, the walls, seats and tables of his dining chamber were washed with rosewater, and rose petals strewn on the floor ensured that every footstep would linger on the nostrils. Courtyard fountains cascaded rose-scented water. Servants washed the hands and feet of guests with scented water, and anointed their heads with fragrant oil. Concealed panels set into the ceiling rotated on command, scattering yet more rose petals on the diners below. Powdered labdanum, storax and nard rubbed into the feathers of white doves released above the diners showered scented particles into the air as

---

[7] For a more poetic version, see William Shakespeare's 'Antony and Cleopatra' Act 2 Scene 2.

[8] Rimmel, E. 1865 *The Book of Perfumes* London, Chapman & Hall, p 85 (Edition 5 (1867)) available as an eBook: http://archive.org/stream/bookofperfumes00rimm#page/n3/mode/2up.

the birds lapped the chamber.[9] During the banquet, slaves carried alabaster bowls and golden bottles of warm water,

> 'and soap well mixed with oily juice of lilies, and poured o'er the hands as much warm water as the guests wish'd. And then they gave them towels of the finest linen, beautifully wrought and fragrant ointments of ambrosial smell, and garlands of the flow'ring violet'.[10]

Marc Antony served a strongly aromatic variety of wine at his table, but if its bouquet was in any way deficient it was fortified with incense resins. Greco-Roman culture believed that to smell nicely on the inside was as important as to smell sweetly on the outside. Meats of all kinds were prepared with aromatic herbs and spices, and served with balsamic wines and vinegars. Banquets in the great houses outdid one another in terms of spices and scented desserts, until the dining set forgot the smell of plain food.

Every aspect of ancient life was accompanied by an exuberant, not to say extravagant, use of incense and fine smells. So extravagant, that at the funeral of his wife Poppaea, the Roman Emperor Nero burned as much incense as all of Arabia could produce in 10 years. She herself was not cremated, as was the custom of the day, but embalmed with the sweetest resins and placed in a fragrant mausoleum. A triumphal entry into Rome by a victorious Consul was considered derisory if piles of incense and gallons of rosewater and scented oils were not lavished on the victor. And each triumphal entry had to be bigger, and more lavish, than the last.

What Cleopatra and Marc Antony sought through their use of scents and perfumes was to create their own odour identities — perfumed personas, as it were — out of which would come the status and attribution each sought to establish in the other. Their carefully crafted scented envelopes revolved around sex and seduction; important horses

---

[9] Donato, G., and Seefried, M. 1989 *The Fragrant Past. Perfumes of Cleopatra and Julius Caesar.* Emory University Museum of Art and Archaeology, Atlanta.
[10] Rimmel p 86.

that have carried perfumed jockeys from the dawn of history. They are not the only horses in the perfume race, however. Some 1,500 years before Cleopatra, the newly widowed Queen Hatshepsut found herself ruler of Egypt but, without a husband at her side, she couldn't become Pharaoh in her own right. She sought alliance with the powerful temple priests, whose appetite for incense was insatiable. Each day they burned frankincense at dawn, myrrh at noon, and 'kyphi' at dusk. Kyphi was the holiest of all the fragrances; 16 scented materials mixed together and prepared while sacred texts were recited. Hatshepsut announced a fleet of five ships would be built to go on a dangerous expedition to the Land of Punt, charged with bringing back as much frankincense, myrrh, cinnamon, cassia, cedarwood, sandalwood, sweet cane, nard, cardamom, dried juniper berries, styrax, labdanum and mastic resins as they could carry. The ships returned piled high with sweet smells and Queen Hatshepsut proclaimed herself the first female Pharaoh — the first female King to gain power through scent! The risk Queen Hatshepsut took to order the construction of five ships and make a voyage of exploration through unknown waters to procure scents was great, but so too was her prize.

Before being proclaimed Pharaoh, the temple priests anointed Queen Hatshepsut with scented oil in a secret ceremony, telling the people that the act of anointment had been carried out by the gods, so endowing the ceremony with greater and mystical significance. All Egyptian Kings were anointed in this way (Figure 1.1).

The symbolism lives on to this day, but no longer in Egypt. All British monarchs formally receive their warrant to rule when they're anointed with scented coronation oil in a service conducted by the Christian church; indeed the word 'Christian' originates from the Greek 'Khristos', which means 'anointed'. The use of oil for anointing British Kings and Queens can be traced back to God's instructions to Moses on Mount Sinai to prepare an oil of myrrh, cassia, cinnamon, and sweet cane for use by temple officials during worship.[11] The specific ingredients of British coronation oil date back to

---

[11] Book of Exodus 30: v 23–25.

Figure 1.1

Anoinment of a Pharaoh with coronation oil. Redrawn from *Jewish Encyclopedia* 1906.

at least the coronation of Richard II in 1377, and probably much earlier, and contains essential oils of orange, jasmine, rosewater, cinnamon, musk, ambergris, and civet together with an oil made from pressed seeds of the Himalayan drumstick tree known as 'oil of ben'. Anointing oil has been used in the coronations of all British monarchs since at least Richard's time, conferring the right to rule when applied to the monarch's head, hands, and breast. As with the Pharaohs, the act of anointment is the most secretive part of the coronation ceremony, conducted under a heavily draped canopy, out of view of those invited to witness the ceremony. The bombing of the Deanery of Westminster Abbey in London during World War II destroyed an ancient vial of oil that had been used for countless coronations. Fortunately, the recipe had been preserved, and a new batch of oil was made up for the coronation of Queen Elizabeth II in 1953. That new vial will likely be used for many coronations to come.

* * *

We can't switch off our sense of smell; we use it with every breath we take. In an 80-year lifespan we take about 673 million breaths, breathing in 360 million litres (80 million gallons) of air. If we had a single grain of rice for every breath we took, we would finish up with a 15 tonne rice mountain — sufficient to feed a quarter of a million people! Each breath brings some chemical message about the outside world into our bodies, whether we are awake or asleep. We recognize the familiar smells of home, the seaside or a flower garden; we know when we're in a perfume shop, a tea factory, a beauty salon, an abattoir, a sewer, or a burning house. We *expect* the world to smell, yet we *rely* upon smells for so very little. Obviously, recognizing the smell of your house burning down or rejecting some decomposing meat you're about to eat might save your life, but for the most part it's the evidence of our eyes and ears that keeps us in touch with the things in life that influence our survival. This perspective favours a Western experience and I readily acknowledge that the sense of smell has more survival relevance for humans living closer to nature than it does to us of the urbanized variety. But even amongst tribal groups most recently known to anthropology, it's the eyes and ears that dominate.

During the course of human evolution many anatomical features of our primate past have been lost; the monkey's tail, for example, is long gone and only the tiniest rudiment remains at the base of the spine. Is it not puzzling that evolution hasn't eliminated the sense of smell in our species — the one species of mammal not using smells in every aspect of their lives — or at least reduced it to a remnant of what it once was? Because the sense of smell is very much alive and well suggests it has a function beyond just saving us from eating rotten meat. The fact that we can recognize over one trillion different smells says that it provides something of what evolutionary biologists call 'adaptive fitness', enabling us to glean a sliver of genetic advantage to contribute to the evolutionary journey of humankind.

* * *

Humans are, frankly, besotted with smells. The scale of the scent and perfume market is astounding. The forecast international sales of fragrances and perfumes is predicted by Global Industry Analysts Inc. to top US$45 billion by 2018[12] — a sum roughly equivalent to the gross domestic product (GDP) of Kenya. In 2010, the fragrance market in the USA alone was worth over US$5 billion. Over 1,000 new fragrances are launched onto the world market each year, and every celebrity who lays claim to the title apparently has a fragrance or two named after them. According to Euromonitor International, the market intelligence company that keeps an eye on such things, in 2011 US$117.8 million was spent on just the five top-selling celebrity perfumes:

1. Elizabeth Taylor's 'White Diamonds' (US$54.9 million) (actress)
2. Derek Jeter's 'Driven' (US$21.1 million) (baseball player)
3. Sean Combs' 'Unforgivable' (US$17.2 million) (actor, rapper)
4. Antonio Banderas' 'Antonio' (US$13.4 million) (actor, film producer, singer)
5. Jennifer Lopez's 'Glow' (US$11.5 million) (actress)

* * *

Although perfumes are of tremendous importance to us, or we wouldn't spend so much money on them, there's something distinctly odd about how we engage with our sense of smell, setting it quite apart from how we engage with sight. Visually, we do what we can to enhance our appearance in the eyes of others. The hairdresser gives us darker or

---

[12] Global Industry Analysts Inc. www.strategyr.com.

lighter hair, even longer or shorter hair and in a myriad of different styles. The beauty salon paints our faces, reddens our lips and sculpts our eyebrows. Dentists straighten our teeth, making them even and shiny white. If we've the stomach for it, and increasing numbers of us do, the surgeon's knife gives us larger or smaller breasts, a straighter nose, a facelift or gives us the means to avoid putting on weight. The objective is always to enhance what we look like. Fashion doesn't disguise the fact that we're human; it doesn't try to make us look like musk deer, civets or beavers. The clothes we wear accentuate the human form, making the best of our good features while hiding the rest. Look at fashions on the catwalks of Paris or Rome; some are no more than a few scraps of diaphanous dalliance, hiding practically nothing and by no means disguising that what's beneath is truly human.

With our olfactory image we do the exact opposite! We do whatever it takes to smell not of humans, but to smell of something else. Cleopatra and Marc Antony bathed in scented water before being set upon by slaves who scraped their skins with bronze scrapers to remove all traces of residual natural smell. The elaborate preparations they made to remove their natural body smells was necessary because their bodies were richly gilded with glands pumping out scents that cry out 'smell me! smell me!', just as do yours and mine. Beyond doubt, humans are the smelliest of the primates. The follicle of every hair, including those of the tiny downy hairs that cover the parts of the body generally thought of as hairless, is equipped with a gland secreting substances onto the surface of the skin that give us our human smell. These are the 'apocrine' glands, named on account of a technicality about how their secretion is made. They produce large quantities of a fluid that feeds the trillions of bacteria living on the skin, particularly in places that are kept warm and moist, and we have some truly outstanding aggregations of them.

The principal aggregations of apocrine glands are in the armpits, where the density of apocrine glands is so high we can regard them as scent-producing organs. Humans have larger armpit organs than gorillas and chimpanzees — our two closest biological relatives — both in absolute and relative terms. The fierce smell assaulting your nostrils when you stand close to a gorilla or a chimpanzee in a zoo is of the secretions of the apocrine glands, upon which bacteria have acted and broken them down into goat-like, rancid-smelling fatty acids. You'd smell even worse than a gorilla were you to neglect your hygiene, because of the greater number and density of scent glands you have in your armpits. Fortunately for those around you, you can do something about it.

Armpits, or 'axillae' (singular 'axilla') to use their correct anatomical name, pose an evolutionary puzzle. Their function is to produce scent; that's what they do and that's all they do. They have no other purpose. Skin scent glands in mammals are common, being used for marking the environment with scented secretions, but we don't apply our armpits to anything in the environment around us. The scent stays where it's produced. In civilized society, we spend much time and mountains of money denying them their biological function. There is no other part of the human body whose function we so assiduously try to extirpate. On the face of it, axillae seem puzzling evolutionary fellow-travellers, but, as we'll see, we owe them our humanity.

Not content with having expunged their natural smells, Cleopatra and Mark Antony perfumed their bodies with unguents of non-human origins, including with paste scraped from the anal pouches of rank-smelling civet cats and from the penis glands of rutting male musk deer. You and I do much the same thing each time we shower and use scented soap — though unless money is no object, your soap will likely contain cheaper ingredients. Next, we pile on deodorants, toilet waters, fragrant aftershaves, and perfumes, all of which contain animal and plant smells. The best of them includes civet, musk-deer or beaver scent gland secretions, and may also contain hints of mammalian urine and faeces. If the perfume is good you don't consciously notice these smells.

Scents and perfumes are present wherever we go and whatever we do; non-scented soap, scent-free toilet paper, unscented paper handkerchiefs, and even unscented garbage bags are hard to find; incense accompanies worship in most of the world's great religions; in various parts of the world visitors are greeted with garlands of fragrant flowers (if not their host's armpits!); cleaning products are scented with citrus and green fragrances; antiseptics have artificial scents added and in some countries you can still find cologne sprays that will freshen you up at the drop of a penny in a slot. What were Herrick, Huysmans, and the writer of the *Song of Solomon* up to, extolling the beauty of human smell, when Cleopatra, Mark Antony and the rest of us are frantically engaged in ridding ourselves of it? This is the enigma of smell.

Chapter 2

An Evolutionary Perspective

To get to the bottom of the enigma we need to go back in time to the Miocene epoch, to when the apes (gorillas, chimpanzees, orang-utans, human ancestors, and gibbons) were starting to differentiate themselves from all other primates (monkeys, lemurs, tarsiers, and lorises). The Miocene, which lasted from about 20 million years ago to about 5 million years ago, was a time of climate cooling and drying, characterized by declining forests and expanding grasslands. Herbivorous mammals ventured out onto the grasslands and began to thrive on the nutritious grasses, gradually becoming the antecedents of the vast herds of antelope and gazelle that occupy the plains to this day. The ancestors of lions, hunting dogs and leopards evolved with them, so the herbivores acquired escape mechanisms of speed and agility. Quickly the predators evolved even better ways of catching them, and the herbivores retaliated with yet more speed and agility, in a balletic evolutionary arms race that nobody must win outright.

Probably because of diminishing food supplies in the declining forests, our ancestral line forsook the trees to take up life on the grassy plains and adopt a very different lifestyle from that which they knew before. No other apes made this move; our ancestral line alone ventured from the forests to the grassy plains. By the end of the Miocene they had separated from their closest relative (the chimpanzees), and their upright bodies bore witness to much evolutionary change.

The prize for living on the grassy plains was a new and abundant food supply offered by the emerging herds of herbivores. Meat is far more nutritious than plant material, even when raw (cooking didn't happen until about 2 million years ago, well after the end of the Miocene), but animals are far more difficult to capture than plants — particularly fast-moving and agile ones, adapted to escaping from fast, well-armed predators. It's safe to assume our forest-dwelling ancestors would have killed and eaten only whatever small mammals, reptiles, birds, fish, and invertebrates that could be found and killed with two hands, or with a stick or stone. Meat would have been a rare component of the diet, which would have consisted mainly of shoots, buds, fruits, nuts, leaves, and flowers. For a species lacking strong jaws, sharp canine teeth, or slashing claws, and not being particularly fast runners, hunting fleet-of-foot antelopes and gazelles presented a considerable challenge. The challenge was met by a number of family groups coming to live together, so there were always sufficient individuals around to be able to collaborate in hunting groups, each playing his or her part in a communal effort to trap and kill large quarries, and each sharing the spoils of the hunt. The challenge was met, too, by an increase in brain size, ultimately resulting in our ancestors being able to solve problems posed by the careful planning and coordination needed for collaborative hunting. Acquisition of language would have been an early facilitator of successful planning.

Not surprisingly, something else happened as our ancestors exploited their newly found food source. They became carnivores. Their intestines shortened and the caecum, once the key to the digestion of plant cellulose, shrank and gradually shrivelled to become the appendix. In modern humans the appendix lacks its original function and for many people their best day is when they are rid of it. The extensive grasslands, richly stocked with large mammals, provided the evolutionary stimulus for our species to adopt a colonial lifestyle, freeing them from the constant need to keep moving around the forest to find new food, as local and seasonal fruit depletion had previously dictated.

There's a shard of biological evidence from another research realm confirming that our early ancestors established home bases and ceased roaming around as their ancestors had been obliged to do. Humans are the only one of the great apes to have a flea living commensally with them. Only mammals that have nests have fleas specifically adapted to them, because fleas lay their eggs in the host's nest. If there is no warm body for the juvenile fleas to jump onto soon after hatching, they die. Gorillas and chimpanzees build temporary nests anew each night as they roam the forest, precluding them from having their own commensal fleas, though occasional, and transient infestations are picked up from other animals. Small details of evolutionary biology such as the human flea, tell us that our ancestors lived a far more settled existence than did their closest relatives.

Prior to emerging from the forests our ancestors would have lived in family groups, consisting of one adult male, one or two adult females and their various young. This is the type of social organization known as monogamy. We are pretty sure our ancestors were monogamous because palaeontology tells us there was little difference in body size between males and females. By contrast, in polygynous species in which one male monopolizes a harem of females, the difference in size between the sexes is very marked. Fossil pre-humans showed no marked size difference between males and females. As a way of life, monogamy is rare among the wider group of mammals, occurring in only about 3% of all species that have been studied. Among the primates, however, it is a lot more common, with about 25% of species showing it to some degree. But among birds, it is the norm; about 90% of species show it. Humans and their ancestors, however, are the only species of mammal to live more-or-less monogamously, *while at the same time living colonially*. This is a unique, and on the face of it, unlikely combination that necessitated some very significant evolutionary adaptations to allow it to work.

*   *   *

Monogamy and polygyny are at the extreme limits of the range of mating arrangements found in animals and just about everything in between can be found. Some species show polyandry, where one female mates with several males, (e.g. marmoset monkeys from South America), and some are able to switch from one system to another depending on ecological or seasonal factors (e.g. chimpanzees). As already noted, the type of mating system employed by a species is reflected in the differences between males and females, called 'sexual dimorphism'. The sexes in monogamous species generally look quite similar to one another. In monogamous gibbons, for example, sexual dimorphism in size is slight, although the sexes may be differently coloured; in humans the size difference is also slight, with men being on average only a little taller and heavier than women. A fairly recent ancestor of ours, *Australopithecus afarensis* ('Lucy', our 3.2-million-year-old Ethiopian ancestor), showed approximately the same amount of sexual dimorphism in body size as occurs in ourselves. Polygynous species on the other hand, show considerable differences between the sexes. Gorillas and orang-utans are polygynous, with males weighing over twice as much as females. Males of both species monopolize bands of females, defending them from other males with exhibitions of strength, chest beating, loud hooting, and displays of formidable canine teeth. Male chimpanzees aren't polygynous and show little sexual dimorphism in body size, but neither are they monogamous. Their mating system is amazingly variable. They live in large to very large loose bands, in which both sexes are highly promiscuous and engage in frequent, freely offered copulation. Their bands split up and regroup with great fluidity, and generally without hostility.

Whatever the type of social organization a species shows, the objective of reproduction is always the same, namely the leaving of offspring to the next generation. The strategy adopted by monogamous species for handing their genes to the next generation is for both parents to be dedicated to the task of bringing up the young, protecting them from attack, providing them with food and teaching them various life skills. Monogamous species tend to have rather few young during their lifetimes, a possible drawback that's compensated for by a very high survival rate of young. It's necessary that males and females of monogamous species must choose their mates wisely, because so much of their life is dictated by the outcome of perhaps a single mating. By contrast, the genetic strategy of polygynous species is for the fittest, strongest, and largest males to leave as many offspring as possible. In such species there are many fringe-dwelling adult males who are unsuccessful in their challenges to win a mating or two. They will challenge again next breeding season but, unless they are eventually successful, it's possible they may not leave any offspring during their lives. This is in sharp distinction to monogamous species, in which most males will mate and most will leave a small number of offspring.

As far as we can tell from fossil evidence of the sexual dimorphism of body size of the males and females of our ancestral species, our line has been typified by monogamy for most of its history. This doesn't mean that philanderers are a modern invention; rather it means that the commonest type of arrangement has been one male pairing with one female. We also know from study of our ancestors over the last 2 million years that their brains became larger than those of their predecessors, requiring increasing lengths of time during which the infants were dependent upon their parents. After birth, the long, slow juvenile development and dependence on parental support for food, protection, and eventually transfer of knowledge benefitted from the continuing presence of both parents, because an offspring with two attendants stood a better chance of surviving than an offspring with one.

It's axiomatic that if two parents are to remain together on offspring-support duties, some mechanism needs to be in place to deter either partner from wandering off to sow wild oats elsewhere before the offspring becomes independent, otherwise, the whole reproductive exercise becomes pointless. As good parenting is an essential element in ensuring that offspring survive through to the next generation, ways evolved to ensure neither partner left the arrangement simply out of boredom. The male is the member of the pair most likely to leave the partnership because his investment in producing an infant is infinitely less than that of his mate's. All he has to do is to inseminate the female, but she has to grow and carry an embryo for many months, and has to find enough food for two along the way. The spoils of collective hunting are needed more by her than by him. Sticking by him and benefitting from his ability to see off interfering males, who are likely only interested in her, and maybe hostile to her offspring, makes good biological, as well as evolutionary sense. A bond between the sexes is needed of sufficient strength to keep the parents together long enough for the young to mature and become independent.

Because monogamy is so rare among mammals there's been much speculation about how it originated, and why in primates it's so common. Evidence from palaeontology and genetics suggest it arose during the past 20 million years — relatively late in the 60 million years of primate evolution. Christopher Opie and his colleagues have examined a huge data set on primate ecology and reproductive biology, and concluded that the main driver for monogamy was the occurrence of infanticide practiced by dominant males when they took over additional fecund females.[13] Infanticide is the major source of mortality in gorillas, accounting for over one-third of all infant deaths; in langur monkeys it accounts for over two-thirds of infant deaths. Dominant male gorillas kill the young of fecund females they monopolize in order to bring the females back on heat as quickly as possible, so they can insert their genes into the next generation at the expense of a previous male's genetic investment with as little waste of time as possible. Opie and his team conclude that monogamy primarily arose as a means to provide added protection to unweaned young against attacks from other males, and only later did the secondary benefit arise, that of increased infant survival through both parents being involved with rearing the offspring. Bonds need constant reinforcing to keep the partners interested in each other. In humans, reinforcement depends upon the continuous availability of sex, facilitated by a raft of anatomical and physiological features that we'll look at later. The pair-bond did the trick of keeping the pairs together, but communal living presented it with a serious challenge.

*   *   *

By being the only mammal that's both monogamous and communal, *H. sapiens* is a unique primate. Other monogamous primates, such as gibbons, occupy discrete territories, whose borders are vigorously

---

[13] Opie, C., Atkinson, Q.D., Dunbar, R.I.M., et al. 2013 Male infanticide leads to social monogamy in primates. Proceedings of the National Academy of Sciences, USA. 110 (33): 13328–13332.

defended by a cacophony of shouting, calling and general border protection flamboyance. There are some birds that are both monogamous and gregarious, including penguins and crested auklets, but right across the vertebrate world the combination is extremely rare. The problem with being both monogamous and communal is that there are always plenty of males around when a female comes into her fertile period, and therefore there is always a chance that a male in a bonded relationship with a female will be cuckolded.

A characteristic part of the sexual strategy of monogamous species is that there's a period of courtship, when a male pays a female close and constant attention for a longer, or shorter length of time prior to mating. He ensures no other male can get close enough to mate with her. As well as enabling each potential mate to assess the quality of the other, courtship is usually of sufficient duration to provide assurance to the male that the female is not already pregnant, because there's no genetic advantage to a male devoting his life to raising another male's offspring. As an anti-cuckolding device courtship works quite well though, like most biological systems, it is not foolproof. Research into the genetics of fledgling birds, for example, has repeatedly shown that females quite commonly solicit matings outside the pair-bond, gaining the benefits of a 'flashier' male's genes while retaining the protection of a solid and dependable mate to help her bring up the young. The same happens in other species, too, including our own.

Female mammals of just about every species advertise their heat, in order to ensure mating occurs when the likelihood of pregnancy is greatest. This was also true for our distant Miocene ancestors. Female primates commonly develop fleshy, bulbous lobes of sexual skin around the vulva and perineum as they approach their peak of sexual receptivity. In chimpanzees the sexual skin is white, in macaque monkeys it is an orange-red, in gelada baboons it is brick red and in colobus monkeys a light blueish-pink. Associated with the development of sexual skin is the production of a specific sex-attractant scent produced by apocrine glands situated around the vulva, as well as from the vagina itself. A well-known example of the power of female sex smells is the so-called 'bitch-run' in dogs, when a female dog on heat is pursued by many males drawn from far and wide by the attractive sex smell she produces, (and engagingly described in Rabelais' outrageous story of Pantagruel's revenge on a woman who rebuffed his amorous advances by spraying her dress with the urine of a bitch on heat).[14] As the ancestral human line evolved, the genetic advantage of visual and olfactory advertisement of heat gradually became superseded by the benefit of ensuring that only one male would remain sexually involved with one female. This lessened the chance that a partnered female would be inseminated by any male other than her bonded mate.

In early 'hominins', as the exclusively human line is known, the florid visual advertisement disappeared under normal selective pressures favouring the pair-bond, until in modern humans no trace of it is left. Since male and female are kept together by the ready availability of sex throughout the monthly cycle, the absence of visual fertility markers was of little significance to a female's bonded mate and would have had no effect on the pair's reproductive output.

Male primates pay great attention to a female's smell as she comes into heat. Male baboons stay close to females during the fertile phase of their cycles, regularly sniffing their perineal regions. The frequency of investigation reaches a peak immediately before ovulation, when mating occurs. In a monogamous species, extinguishing the smell has an effect similar to the loss of visual cues; of little importance to the members of a bonded pair but of vital importance with respect to the sexual anonymity it brings to the outside world. But unlike visual cues, which gradually disappeared over time, something very different happened to render the olfactory cues meaningless. It came about through the agency of a chance mutation affecting not the nose, but a part of the sense of smell that modern humans have come to live happily without.

*   *   *

---

[14] Rabelais, F. 1653 *Five Books of the Lives, Heroic Deeds and Sayings of Gargantua and his son Pantagruel*. Book 2, Chapter 2 Section 21. Electronic Classics Series, Pennsylvania State University, Hazleton.

The mammalian sense of smell consists of two separate systems, one based on the nose's olfactory membrane, and the other on a structure known as the 'vomeronasal organ', or VNO for short. Mammals utilize both the nose and the VNO to interrogate the scented world and while some species make it clear which system they are using at any particular time, it is not true for all. It is an important sensory tool for just about all mammals. Only a few types have no VNO system; whales and dolphins have no sense of smell at all, nose or VNO. The great apes, humans, and the 'Old World monkeys', (monkeys from Africa and Asia), also lack functional VNOs — but monkeys from the 'New World' (Central and South America) still retain functional organs. With these few exceptions, the VNO can be regarded as a mammalian characteristic. The VNO consists of a pair of blind-ended tubes, embedded in the roof of the mouth in a bone of the hard palate called the 'vomer', lying right underneath the nasal septum — the sheet of cartilage dividing the right and left nostrils. It connects with the mouth cavity *via* a pair of fine canals. In the majority of mammals, including those primates that have it, the VNO acts as a sex smell detector, passing its nerve impulses directly to the part of the brain that controls sexual behaviour. Plenty of experimental work has been carried out on many different kinds of mammals to show that the VNO system must be intact for sex to occur. Sex clearly occurs in humans, the great apes and the Old World monkeys all of which lack functional VNOs, so how do they get along without it?

At a seminal point in primate evolution, shortly before the great apes and the Old World monkeys separated from the ancestral primate stock about 23 million years ago, a random mutation occurred to the genome that prevented the VNO from getting its message out to the brain, effectively crippling it and rendering it without function. The mutation was given the nickname 'ADAM' by its discoverers, which I'll write in upper case letters to distinguish it from the *Adam* of my title. We'll look closely at 'ADAM' in Chapter 6, but suffice it to say that it broke the ancient link between smell and sex.

By decommissioning the VNO, 'ADAM' removed vital information about the stage of a female's ovulation cycle, because males were unable to make meaning of scent signals perceived *via* the VNO. The threat to the pair-bond posed by the close presence of other males seeing visual cues, and smelling scented cues, was now eased. 'ADAM' helped ensure that the pair-bond could withstand the pressures of communal life. Together, the adaptations to visual signalling and VNO scent perception meant that ovulation was privatized; no one knew when a particular female was entering the fertile phase of her cycle. The weakness of the pair-bond's principle trick — of making sex readily available throughout the cycle — was now overcome. Adaptations to the visual and cueing of oestrus now meant that impregnation by the bonded male, rather than some passing Lothario, was more assured, and so the genetic advantages of monogamy were protected, and our ancestors could live communally while maintaining a monogamous lifestyle. Although the mutation occurs in all the apes and Old World monkeys, its real importance lies in the only species amongst them that is *both* monogamous *and* gregarious — that is, humankind. The profound importance of 'ADAM' was that it freed our ancestors from slavish responses to sexual smells, and so in freeing them, made us human.

*　　*　　*

And so we come back to the enigma of why humans are uneasy with the smells of their own bodies. Although 'ADAM' knocked out the VNO, it didn't destroy the ancient neural pathways in the brain formerly associated with sexual smells. As human ancestors gradually evolved from their newly carnivorous ancestors over the next few million years, the first semblances of a smell culture started to emerge, though exactly how the first early or pre-human came to spread some nice smelling substance on his or her body *for the sensory pleasure it gave*, remains a mystery. Perfumes, especially those made from sex attractant scents of animals and plants subconsciously release an awareness of the wearer's animal presence. Over the past quarter of a million years, our unique species of ape came to regard the natural body smell as having a harmful effect on emerging social structures through its ability to telegraph our animal origins. As running water and soap became more common, our forefathers went to increasing lengths to remove their natural body smell, but sensing a certain social nakedness associated with having no smell, replaced it with something appealing to the nostrils. That something didn't speak of human bodies.

The enigma is now resolved. Monogamy and gregariousness could co-exist only when the link between sex and smell was broken; only when the link had been broken could our somewhat puny ancestors find ways to hunt cooperatively to kill the large mammals on the grassy plains of Africa. Without the link being broken, we would not be here today. As societies grew and people came to live at ever-higher densities, constant reminders of our animal origins didn't sit well with emerging social and religious expectations, and so people did what they could to remove their natural smells. Twenty-first century Adam and Eve have made an art form out of removing their natural body smells, scrubbing, shaving, deodorizing and perfuming all parts. As we'll see later, sex must still smell, so amid the need to present a nice public image to the nose, perfumes redolent with animal musk and urine-smelling compounds long known for inviting what Plato called 'carnal indulgence', came into favour. The tension between the animal sensuality of smell and the human intellect, which can soar above the baseness of our animal origins, helps to explain the appeal of so much poetry, literature, and art that extols the hidden meaning of human scent.

Chapter 3

How Smell Works

At this point it's necessary to step aside from the evolutionary story to see how smell works if we're to understand how scents interact with our *internal milieu*, uplifting and fortifying us, or driving us to Plato's concupiscent lusts. To do that we need to go back to when life first appeared on Earth; we don't know where this occurred, though the matter has been the subject of intense debate over the last century and a half. Charles Darwin spoke of a 'warm little pond' as the likely cradle in which some spontaneous chemical activity gave rise to life. Wherever it arose, the earliest identifiable forms of life are primitive bacteria-like microorganisms whose fossilized remains have been found in 3.4-billion-year-old rocks. David Wacey and his team working in Western Australia reported finding spherical and tubular cell forms fossilized in the rocks.[15] These organisms mark the earliest known occurrence of life on Earth. As with all bacteria their single-celled 'bodies' lack a nucleus and for this characteristic are called 'prokaryotes' — the word literally means 'before a kernel'. Humans belong to the huge group of organisms, including fungi, plants and animals, whose cells are equipped with nuclei. Collectively this group is known as 'eukaryotes', or organisms that have 'formed kernels'. They first made their appearance 3 billion years ago (Figure I.1). The nucleus is the place where chromosomes carrying DNA are held.

---

[15]Wacey, D., Kilburn, M.R., Saunders, M., *et al.* 2011 Microfossils of sulphur-metabolising cells in 3.4 billion-year-old Archaean rocks of Western Australia. Nature Geoscience 4: 698–702.

While the number and species richness of the prokaryotes is incalculably high — there may be as many as 10,000 types of bacteria in a single drop of seawater — we know all too little about them. Our familiarity with the living world is heavily based on what we know about the eukaryotes.

A characteristic of bacteria is that they are all very tiny and you need a microscope to see them. Exceptionally, they live in vast colonies that are not only obvious to the naked eye, but form clear geographic features. Not uncommon in shallow, salty tropical lagoons around the world are knobbly, rock-like structures called 'stromatolites'. The word literally means 'layered rocks', because within the metre-high pillars can be seen many thousands of densely packed layers. They are built from countless generations of blue-green bacteria, interspersed with fine silt and sand. Stromatolite formations several thousand years old are not uncommon and fine examples can be seen in Shark Bay in Western Australia. The bacteria building them capture energy from the sunlight using photosynthesis and have played a pivotal role in the evolution of life on Earth, because a waste product from photosynthesis is oxygen. When bacteria first appeared on Earth there was no atmospheric oxygen, but as air-breathing animals didn't appear until almost three billion years later, there was plenty of time for its levels to accumulate.

Although bacteria have no noses as we would recognize them, or any sense of smell as it's commonly understood, it's in their simple cell structure that we find the origins of our ability to perceive smells. In essence, how all animals perceive smells is no different from how bacteria perceive chemicals in the world outside their cell walls. The sense of smell is our most ancient sense, but just how ancient is quite staggering. By comparison, our eyes and ears appeared but an evolutionary instant ago, before shooting to sensory supremacy. But, like the tortoise and the hare, the sense of smell has plodded along through eons of evolutionary time, emerging into the 21st century much the same as when it first appeared. It's a part of our bodies that directly links us to the earliest days of life on Earth.

*   *   *

There's an important physical difference between what humans regard as a smell and what, say, an octopus regards as a smell, but there's little biological difference. It's all to do with the *physical size* of the chemical molecules constituting the smell and much less to do with their chemical structures. To be airborne, a molecule must be small and light; its lightness allows it to be lifted into the air and carried on the wind to our noses. Airborne molecules must weigh no more than the combined weight of 300 atoms of hydrogen gas — the lightest chemical element of all. The weight of one atom of hydrogen gas is called a Dalton, after the English physicist John Dalton; thus 300 Daltons is the limit for a chemical to be airborne. Heavier molecules can't float in air any more than rocks can float on water. Such a constraint doesn't apply in an aquatic environment however, where any water-soluble molecule can diffuse through the water; its rate of travel is influenced by the nature of the chemical, the temperature and salinity of the water, and on currents. Bacteria and aquatic animals can perceive molecules that are large and unable to become airborne, as well as tiny molecules that could easily float in air — for them there's no difference between tasting and smelling. For Adam and Eve, the difference between what the mouth and nose tells them is very significant. Our senses of taste and smell can work independently of each other or, as when you sit down for dinner, jointly.

A bacterium is a single-celled organism capable of carrying out all its biological functions within the confines of a single cell. Unlike multi-celled organisms that have specialized parts of their bodies for different functions, bacteria conduct all activities with the equipment contained inside their single cell. They have no separate sense organs with which to detect changes occurring in their environment, instead using their whole 'body' for the task.

The bacterial cell is a little bag of organic material, bound by a membrane. In bacteria that invade our bodies and make us ill, the membrane is surrounded with a capsule to protect it from the body's immune system, but in free-living forms there is no capsule. The membrane surrounding the organic material is designed to regulate the transport of molecules into and out of the cell, such as food, water, and waste products, as well as facilitating the detection of external

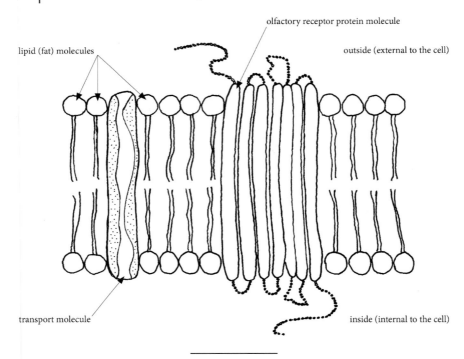

Figure 3.1

Schematic section through a cell wall. A seven-limbed olfactory receptor protein (GPCR) spans the entire thickness. The cell wall is between 20 nanometres and 80 nanometres thick — between one five-thousandth's, and one thousandth's the thickness of a human hair.

chemicals. The cell membrane consists of a sandwich-like double layer of molecules, the outsides of which (the slices of bread) attract water-soluble chemicals, while their intermingled fatty tails (the filling) repel water-soluble chemicals. The double layer of the cell membrane is impermeable to water and prevents substances from passively diffus-ing in or out. The membrane is punctured with special channels and ducts that can open to allow the passage of chemicals. It is also equipped with molecules called 'receptors' that have the function of receiving chemicals on the outside of the cell and transferring informa-tion about them to the inside. These molecules are called 'chemorecep-tors', because what they do is to respond to chemicals (Figure 3.1).

Multi-celled organisms respond to external chemicals in fundamentally the same way as do bacteria. Both have receptor molecules projecting through the cell membrane that bind with odorant molecules on the outside. Both have some quite complicated biochemistry and biophysics to trigger a biochemical chain reaction inside the cell, resulting in signals being sent to whatever structures control the body's orientation — a whip-like flagellum in the case of bacteria, tentacles in octopus, fins, wings, legs in mammals, and so on. In multi-celled animals, a single receptor cell generates a tiny electrical discharge that travels to the central nervous system, or the brain, where the nerve discharge is processed and signals are sent out to the appropriate motor organ. Though there are differences in the nature of the protein molecules between pro- and eukaryotes, they don't affect the final outcome. Despite all evolution's twists and turns over the last 3.4 billion years, and the changes that have occurred since life's remotest ancestors left Mr Darwin's 'warm little pond', we humans smell the world through equipment that became functional when life on Earth first arose.

∗   ∗   ∗

Multi-celled organisms have the luxury of having blocks of cells able to specialize as chemical sense organs. There are obvious advantages in this as they can be positioned on the organism where they are most needed. In jellyfish and parasitic worms, for example, they lie close to the mouth; in spiders they are on the legs; in insects they are on their antennae; in animals with backbones they are on the head, associated with other sense organs and close to the brain. In humans and other air-breathing animals, they are housed in the nose.

The heart of every olfactory sense organ is the chemoreceptor molecule. Each chemoreceptor is a protein molecule called a 'G-protein coupled receptor', or GPCR for short (Figure 3.1). This rather unwieldy name comes from the molecule's ability to interact with a protein inside the cell called 'guanosine triphosphate' — the 'G-protein' of the receptor's name. The coupling of an odorant molecule outside the cell with the GPCR inserted through the membrane is the first step in the complex biochemical cascade that takes place inside the cell. GPCRs are long molecules consisting of twisted chains of amino acids, the sequences of which enable them to discriminate between one chemical and another and, ultimately, let you discriminate between different smells. The process of converting the presence of a chemical outside the cell into a nerve response is called 'transduction'. The GPCR and the so-called 'G-protein transduction system' evolved about one billion years ago, long before air-breathing animals appeared on Earth (see Figure I.1).

*   *   *

A major breakthrough in understanding how the sense of smell works came in the early 1990s, when Linda Buck in Seattle and Richard Axel in New York published a paper identifying a suite of about 1,000 genes in mice that expressed themselves in the olfactory membrane as olfactory receptors.[16] The words 'expressed themselves' mean that the protein products of the duplication of these genes occur only in the olfactory membrane — the tiny patch of sensitive tissue lying at the top of the nasal cavity, right underneath where the pads of your glasses touch the sides of your nose, and covering a combined area the size of a postage stamp. The discovery of olfactory receptor genes had been eagerly awaited, giving olfactory research a major shot in the arm, because researchers at last had something they could work on that reflected 3.4 billion years of evolution. The olfactory sub-genome, as the olfactory genes collectively are known, is an evolutionary palimpsest, much like a recorded manuscript of life on Earth that has been erased, modified, erased again and written over many times. The power of modern molecular genetics is that it allows what was written before, and subsequently erased, to be deciphered. The olfactory receptor gene family remains the largest family of genes yet identified and accounts for about 3%, or about 750–800 genes, of the entire human genome. For their work, Buck and Axel won the 1994 Nobel Prize for Medicine and Physiology, so fundamental was its significance.

---

[16] Buck, L., and Axel, R. 1991 A novel multigene family may encode odor recognition: a molecular basis for odor recognition. Cell 65: 175–187.

Following Buck and Axel's work, many studies developed and refined the story, and the picture is now developing quite quickly. The genes for sensing chemicals in the outside world are arranged in six so-called 'multi-gene' families. Four of them encode receptors for smell and two for taste. A multi-gene family arises when there has been rapid duplication of a gene, resulting in multiple genes that subsequently diverge through mutation, though they retain much in common with their fellows. Frequently they are located on the same chromosome as the one which they duplicate, though gradually they disperse to other chromosomes, due to the manner in which chromosomes pair and break as they separate, with bits of one becoming stuck to another.[17] Multi-gene families are further classified into gene families, when 40% of their amino acid sequences are in common, and gene sub-families, when 60% of their amino acid sequences are in common.

All genes are put into two major groups based on the degree of similarity of their sequences of amino acids, and it appears that one class of genes, called Class I, is found only in fish, while all four-limbed animals (amphibians, reptiles, birds, and mammals, collectively called 'tetrapods') have sequences that put them into Class II. Vertebrates have several families of genes within Class II and lots of sub-families. Most species have sub-families specific to them; humans have sub-families not shared with chimpanzees and they have some not shared with us. Dogs have over 30 sub-families not shared with any other species, and so on.

*   *   *

A puzzling feature of the olfactory genome is that it contains many genes that are non-functional and inactive, meaning that they're not able to express a specific protein during cell division. Non-functional genes, or 'pseudogenes' as they're called, are surprisingly common. We humans, together with chimpanzees, hold some sort of record on the pseudogene front, because over half our olfactory receptor genes

---

[17]Nei, M., Niimura, Y., and Nozawa, M. 2008 The evolution of animal chemosensory receptor gene repertoires: roles of chance and necessity. Nature Reviews Genetics 9: 951–963.

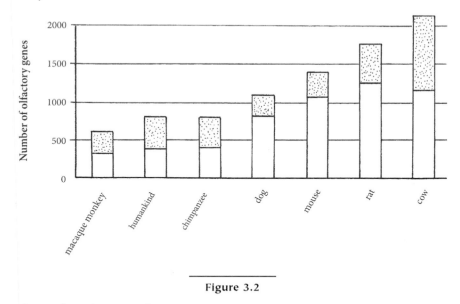

Figure 3.2

The number of active and inactive genes in several species of mammals. Clear bars denote active genes; stippled bars denote inactive genes (pseudogenes).

are represented by non-functional dummies (Figure 3.2). Comparing the whole human genome with that of other mammals, inactive chemoreception genes are over-represented, suggesting this reflects a uniquely human condition.[18]

Each non-functional gene resembles a formerly functional one, now rendered non-functional through the incorporation of protein material that disrupts the gene's ability to replicate itself. Because of their shared ancestry with functional genes in other species, inactive genes contain lots of evolutionary information of immense value to evolutionary biologists.

---

[18] The inactivity of some immune system genes appears — counter-intuitively — to provide protection against severe sepsis. Inactivity occurred after humans had split from chimpanzees. Wang, X., Grus, W.E., and Zhang, J. 2006 Gene losses during human evolution. PLoS Biology 4(3) e52 doi: 10.1371/journal.pbio.0040052.

The discovery that the majority of our olfactory genes are non-functional raises questions about just how good is our olfactory ability. The gene story is still a bit sketchy because we do not yet have enough fully worked genomes upon which to base a definitive evolutionary tale, but what we already know is intriguing. Much work in this area has been conducted by Yoav Gilad and his colleagues at the Weizmann Institute in Rehovot, Israel, and later at the University of Chicago.[19, 20] By genetically sequencing 50 human olfactory receptor genes from chimpanzees, gorillas, orang-utans, and macaque monkeys, and comparing them with dogs and rodents, Gilad showed that all the primates have been losing functional genes at a faster rate than dogs and rodents, and within the primates, humans have been losing them fastest of all. On the basis that more functional smelling genes would make for a better smeller, (dogs have only about half the number of inactive smelling genes as we do) humans should have a poor sense of smell. This prediction might tempt us to say that our reliance on sight and hearing is a consequence of the loss of smelling genes. But in tests comparing human ability to detecting smells with other animals, there's mounting evidence showing our sensitivity to test odours is at least equal to that of other animals and, in some instances, surpasses it. And in picking the odd-man-out in trios of smells, our discriminatory ability seems almost limitless.

Are human olfactory receptor inactive genes simply the remains of genes that were once functional and are now consigned to the evolutionary waste bin? Gilad and his colleagues suggest that while there may be some truth in this, it's far from the whole story. They have shown that about 80 human olfactory inactive genes belong to a specific sub-family that has recently expanded from three inactive genes occurring only in apes. In other words, the 80 inactive genes are of relatively *recent* origin, and are directly descended from the common ancestor from which both humans and apes evolved, between

[19] Gilad, Y., Man, O., Pääbo, S. *et al.* 2003 Human specific loss of olfactory receptor genes. Proceedings of the National Academy of Science, USA 100(6): 3324–3327.
[20] Gilad, Y. 2009 Olfactory receptor genes: human loss during evolution. In: Squire LR (ed). *Encyclopedia of Neuroscience*, Academic Press, Oxford vol. 7, pp 157–161.

four and eight million years ago. While 80 inactive genes represents a small fraction of the whole inactive gene population, this observation shows much more remains to be learned about inactive genes and why they are retained in the genome at all.

There's evidence that the loss of functional olfactory receptor genes through acquisition of inactive genes in humans isn't something that occurred only in the past, however recent or ancient, but is ongoing. Idan Menashe and his colleagues at the Weizmann Institute have shown that 70% of the hundreds of human inactive genes carry only one coding fault. As non-functional genes acquire faults in relation to the length of time they have lost their function, the acquisition of so few faults suggests they lost their functionality quite recently.[21] Menashe examined the genotypes of 86 Americans of European origin and an equal number of people of African origin, and found Americans of European origin carried twice as many faults as the group of Americans of African origin. While the samples are small, and the work not sufficiently comprehensive to allow definitive conclusions to be drawn, it indicates that surprisingly large differences have occurred in the very short period of time that human beings have been human, and that distinct sub-populations of human olfactory receptors exist in humankind today.

Putting all these scraps of evidence together, a fascinating picture of the human olfactory sense starts to emerge. Firstly, between a half and two-thirds of our olfactory receptor genes don't work — we have only around 345 that do — and while primates have a lower proportion of non-functional genes than we humans, rodents, dogs, cows, opossums, chickens, toads, and fish have much lower proportions still. Secondly, the process of the loss of functional olfactory receptor genes may be ongoing, and various parts of the human population may carry different proportions of gene faults. Thirdly, the loss of functional olfactory genes appears to be a gradual process and is probably ongoing. Fourthly, some human olfactory inactive genes appear to have been acquired quite recently from the common

[21] Menashe, I., Man, O., Lancet, D. *et al.* 2003 Different noses for different people. Nature Genetics 34: 143–144.

ancestral stock, our line shared with the apes, leaving open the possibility that inactive genes may not be quite as inactive as is generally thought. And finally, it appears that each of us has a highly personalized set of olfactory receptors that affects how each of us perceives smells.[22] About a third of my olfactory receptors are different from yours, meaning that while both of us will agree that the flower we are smelling is a rose, the actual receptors each of us uses to make this determination will not be identical.

The evolution of olfactory receptor genes is being studied in a number of the world's leading genetic laboratories and while it's hard to give an unequivocal account of how receptors have evolved from the time the first four-limbed beast lumbered ashore, some promising leads are emerging. Examination of the coelacanth, the most primitive four-limbed fish alive today that hides in the deep waters off Madagascar, reveals the presence of both Class I (fish) and Class II (tetrapod) genes, but most of the Class II genes are not functional.[23] By contrast there are no *non*-functional Class I genes, telling us that despite its four 'legs', it remains very much a fish. In all fish, including the coelacanth, Class I olfactory receptor genes are clustered on just one chromosome — chromosome number 11. This is a rather special chromosome because about 40% of Adam's and Eve's olfactory receptor genes are carried on it. The other 60% are carried on every other chromosome except number 20 and the female sex chromosome (the 'X' chromosome). Chromosome 11 may have been the original site for smelling genes, way back when animals with backbones started to make their appearance on Earth.

We know from the study of fossils that tetrapods evolved in the Devonian epoch, about 400 million years ago, from relatives of the coelacanth. As we don't have access to body tissues of these ancient creatures, and regrettably can't conduct any genetic analyses, we can only guess at what olfactory receptor genes they were carrying. It's

---

[22] Mainland, J., *et al.* 2013 The missense of smell: functional variability in the human odorant receptor repertoire. Nature Neuroscience 17(1): 114–120.

[23] Hoover, K.C. 2010 Smell with inspiration: the evolutionary significance of olfaction. Yearbook of Physical Anthropology 53: 63–74.

tempting to speculate that they, like their present-day relatives, had both classes of genes and that the Class II genes were functional and that they went on to provide the repertoire of functional olfactory receptor genes for all the air-breathing tetrapods that came after them, but I'll resist the temptation until we know more about inactive genes.

*   *   *

Ancient relatives of amphibious newts and salamanders were the first to live on dry land, though to this day all must return to water or otherwise provide their embryos with water in which to develop. Amphibians breathe air through nostrils and have lungs, but when they are in water, oxygen diffuses across their skin. Amphibians possess both Class I and Class II olfactory receptor genes, with many more Class II receptors than Class I. Anatomical studies on the Indian clawed toad *Xenopus,* much studied four decades ago when this luckless creature was found to be highly responsive to human pregnancy hormones and was used in early pregnancy tests, reveal that Class I receptors — the fishy ones — are expressed in small lateral sacs in the toad's nose in a place ideally suited for sensing the aquatic, rather than the terrestrial environment. Class II receptors are expressed in the olfactory membrane in the toad's nose, just as they are in yours and mine.

In a comprehensive study of olfactory receptors of aquatic and terrestrial animals, Joachim Freitag and his colleagues in Stuttgart examined a common species of dolphin inhabiting temperate and tropical waters, because the established wisdom is that dolphins and whales have no use for an olfactory system. Freitag could find only Class II receptor types in the dolphin's genome and all were represented by non-functional inactive genes.[24] Furthermore, they showed quite high numbers of genetic defects in the inactive genes, suggesting the absence of any positive selection pressure over a long period of time. The diversity of gene families and sub-families in the striped

---

[24] Freitag, J., Ludwig, G., Andreine, I. *et al.* 1998 Olfactory receptors in aquatic and terrestrial vertebrates. Journal of Comparative Physiology A 183: 635–650.

dolphin resembled the diversity of receptor genes in other mammalian species, but their number was significantly reduced. Dolphins evolved during the Eocene, shortly after the extinction of the dinosaurs, about 50 million years ago. The earliest known fossil whale, from what is now northern Pakistan, was a dog-sized carnivore that most likely sought its food in the Tethys Ocean, for at that time the land was becoming drier and warmer, making traditional food harder to find. As their terrestrial relatives evolved to a plant-eating way of life and went on to evolve into cattle, horses, and all the other large herbivores, the ancestors of whales retained their carnivorous lifestyle. Good eyesight and the development of extremely acute hearing rendered their sense of smell redundant.

*   *   *

There's one last matter that needs to be discussed before we can take a look at how animals pick up smells. It concerns precisely *how* a particular odorant molecule is able to trigger the biochemical chain reaction inside the cell, such that we perceive and identify the smell of baking bread when we enter a bakery, and not the smell of, say, decaying meat. Here we face one of the most enduring and impenetrable problems in olfactory science; it's impossible to predict the smell of a chemical molecule, or to predict the chemical structure of any particular smell. Long ago it was discovered how our eyes perceive an image, and create it in colour, too, and how our ears are able to discriminate the tiniest changes in pitch and timbre. But a complete understanding of how our oldest sense works still largely eludes us.

In our noses we have about 345 receptor types; if each smell was represented by one type of receptor we'd be able to perceive only 345 smells. Clearly that's not the case as we can recognize infinitely more, so somehow the system must work through various patterns of receptor stimulation. For centuries it was assumed that smells exhibited physical properties to which the nose responded; fierce smells were thought of as jagged and sharp, tearing at the nasal membranes with rough edges and spikes, while sweet smells were smooth and rounded, soothing the nose as they were inhaled. Eventually these ideas gave way to a classification of smells into six categories reminiscent of primary colours, *viz*: floral, putrid, fruity, spicy, resinous, and burnt, from which palette all other smells could be mixed. It was quickly found that this didn't work, so the simple idea of primary smells forming the basis of all others languished.

The next theory to gain acceptance was also based on the three-dimensional shape of an odorant molecule. The theory held that two or more molecules with the same physical shape, irrespective of their chemical composition, would smell the same. This came to be known as the 'lock-and-key' theory, working on the principle that just as the pattern of cuts on a key will be able to open only one lock, so only a particular shape of smell molecule would be able to release a particular perception. Seven basic ingredients each with a distinctive shape — camphor, pungent, ethereal, floral, peppermint, musk, and putrid — were thought to represent the range of molecular shapes in nature. The lock-and-key theory still holds adherents, despite clear indications that it's unable to explain the whole story.

Luca Turin, a British scientist who has done much to further our understanding of how smells smell, demonstrated with an elegant experiment that the lock-and-key theory doesn't explain how all smells are perceived. He used two molecules with very similar shapes, but very different chemical structures. Camphene has a strong aromatic smell often encountered in Indian cuisine, while decaborane, smells pungently of sulphur. Each molecule has a central spine of ten atoms; in camphene the spine's made of carbon atoms, while in decaborane it's made of boron atoms. The lock-and-key theory predicts that their very similar shapes should result in similar smells, but very obviously this isn't so. The theory further predicts that sulphur-smelling compounds should have a bulbous shape, since sulphurous compounds all have a bulbous end, but decaborane has no such shape. Turin began to think about what other characteristics a molecule displays that might explain why these two similarly shaped molecules have such different smells.

He dusted off observations made back in the 1920s showing that molecules vibrate when infrared light passes through them, and that the speed with which they vibrate, called the vibrational frequency or spectrum, is a specific characteristic of them, something like a chemical fingerprint. Michael Dyson, the British chemist who worked on vibrational spectra in the 1920s, hinted that the vibration of molecules might help explain why they smelled as they do, though he had a limited understanding of olfactory receptors. His theory gained support in the 1950s when equipment for specialized far-infrared analysis became more widely available. Together with a Canadian chemist named R.H. Wright, Dyson proposed that different frequencies of vibrations stimulated different parts of the nose's receptive membrane, rather as the sound of a chord played on the piano is made up from the individual vibrations of the various strings hit by the keyboard's hammers. The smell of a particular substance was the combination of the vibrations of each of its constituent parts. They envisaged the olfactory membrane as a kind of spectrometer — working rather like a glass prism separating the parts of a mixture into its components. The theory made sense and had many adherents, but there was a problem.

Smell molecules sometimes come in left- and right-handed versions, called 'enantiomers', each a mirror image of the other. You can envisage them as your two hands; your left hand is a mirror image of your right. Both left- and right-handed versions have the same chemical structure and the same molecular weight. They also produce the same vibrational spectrum. The majority of them smell the same but, significantly, many don't. It's been estimated that 64% of enantiomers of odorant molecules share the same smell, and 17% have different smells (this percentage may yet rise as more molecules are tested). While enantiomers share the same chemical structure and vibrational spectrum, they can't fit into the same 'keyhole' any more than you can wear a well-cut glove on either hand.

Luca Turin approached this problem by looking again at molecular vibrations, and considering how a molecule may be able to stimulate a receptor. He figured that if the nose could work as a spectroscope, as Dyson and Wright had hinted, with different receptors responding to different vibrational spectra, the pattern of which receptors are activated could account for the perceived smells. He's written engagingly about his quest for understanding in a little book titled *The Secret of Scent*[25], which is well worth reading. Turin has proposed that a combination of the shape theory and molecular vibration might be responsible. He further proposed that molecules trigger the receptor through a process called 'electron tunnelling'.

Imagine the receptor as being 'U'-shaped, with one arm of the U receiving the odorant molecule from outside, while the other arm receives electrical charges, i.e. electrons, from the odorant molecule. Receipt of charge initiates a biochemical cascade inside the cell. Electrons swarming around a molecule are like a herd of cats; they don't cling very tightly to their parent, and exhibit a tendency to escape. Once a smell molecule has reached the receptor, electrons will cross the gap between the two arms and disrupt the receptor's electrical charge. Energy is lost to the smell molecule because of the loss of electrons from it, exciting it to vibrate. As increasing numbers of electrons tunnel through the gap between the arms, the biological cascade inside the receptor is initiated.

Testing of Turin's electron tunneling theory is still in its infancy. Many details need to be filled in before we can be sure how it is that left-and right-handed molecules sometimes smell alike and sometimes differently, and what role is played by a smell molecule's shape. Turin and his colleagues have made a start by examining the effects of substituting the hydrogen atoms in musk molecules with atoms of deuterium — so-called 'heavy hydrogen', because of the existence of an additional neutron to the atom.[26] Neutrons don't affect the electrical charge of the hydrogen atom; they only increase its mass, but in every respect, other than they have non-identical vibrational spectra, the molecules are the same. He found that people could discriminate between the smells of normal musk and deuteriated musk. All the members of a panel of sniffers could clearly detect the difference between the animal smell of normal musk and the burnt,

---

[25] Turin, L. *The Secret of Scent*. Faber and Faber, London.
[26] Gane, S., Georganakis, D., Maniata, K. *et al.* 2013 Molecular vibration-sensing component in human olfaction. PLoS ONE 8(1):e55708 doi:10.1371/journal. pbio.0020005.

nutty smell of deuteriated musk. In another experiment he showed that fruit flies were able to discriminate between normal and deuteriated acetophenone, both of which smell identically (of honeysuckle) to humans, in which the only difference between the molecules is that their vibrational spectra are not identical.[27] In both experiments Turin concludes that what is being detected is solely the difference between the molecular vibrational spectra of the closely related odorants, and not their specific chemical composition. Much more work needs to be done before we have a clear understanding of how things smell as they do, and it may turn out that one theory will probably not fit all. Turin's ideas take us a step further towards that understanding.

*   *   *

Our excursion into the biology, biophysics, cell and molecular biology, and genetics of the cell indicates that the manner in which humans perceive smells is not fundamentally different from how bacteria perceive chemicals in their watery world, despite all the anatomical differences between them and us. The fundamental element of the sense of smell is the olfactory receptor molecule, the GPRC, which links the outside world with the inside of the cell; the GPRC switches on a transduction chain in which messenger molecules, excited by chemical signals, open channels in the cell membrane, allowing electrical depolarization to occur, sending nerve impulses to the central nervous system. Bacteria have to combine receiving and responding within the single cell, so no nerve impulses are needed. In mammals, the nerve impulses generated by the receptors, head for the brain and are translated into a form that lets you take action, from running away, to pursuing its source as food, or to putting a name to a scent. Turin's work is intellectually pleasing because it supports the idea that the sense of smell is a spectral system, akin to sight and hearing. In each case a huge array of outputs can be generated from a small

---

[27] Franco, M.I., Turin, L., Mertshin, A. *et al.* 2011 Molecular vibration component in *Drosophila melanogaster* olfaction. Proceedings of the National Academy of Sciences USA, 108 (9):3797–3802.

number of receptors. Just as we are able to combine the inputs from only three types of visual receptor into somewhere between 2.3 and 7.5 million colours, and combine the inputs from our acoustic receptors in our ears to provide 340,000 distinguishable tones so, it appears, we can discriminate over one trillion different smells from our 345 chemical receptors.

Chapter *4*

To Catch a Whiff

Leaving aside the unfinished story of how smelly molecules trigger chemoreceptors, and entering the world of higher animals, the range of devices used for smelling is almost as varied as the animal kingdom itself. Among air-breathing animals, noses are dominant facial features. Snouts and noses vary enormously among animals, with species depending most on their noses having the most obvious sensory equipment. We humans have prominent, though not overpowering noses that come in a variety of shapes and forms; aquiline, lofty, straight, wide-nostrilled, hawkish, hooked, snub, flat, and upturned, they are as varied as the faces they grace. Figure 4.1 shows a whimsical compilation of types of noses.

During the 19th century the 'science' of 'nasology' gained pseudo-scientific popularity. It purported to determine a person's personality from the shape of their nose, and was almost as popular as 'phrenology' — the reading of the bumps on a person's cranium. Both vied for top prize among the popular pseudosciences of the 19th century. Ideas about a person's nose and moral character have been around for centuries; Laurence Sterne's 1750s tale *Tristram Shandy* introduces the reader to one Hafen Slawkenbergius who, possessed of a mighty nose, sent the good burghers of Strassbourg into a frenzy as he made his way through their town to find his sweetheart, for a nose such as his had never before been seen. Slawkenbergius had, Sterne tells us, studied the relationship between a person's nose size, shape, and their character, and was an authority on the matter. In the mid-19th century a canny Englishman named Eden Warwick, *aka* George Jabet, clearly fuelled by

Figure 4.1

College of human nose types. Sixteenth century woodcut by an unknown Florentine artist. Redrawn from Warwick E., 1848 *Nasology, or Hints Towards a Classification of Noses*. Richard Bentley, London.

Sterne's creativity, wrote *Nasology; or Hints Towards a Classification of Noses*, an enduring work that was reprinted a number of times during his lifetime.[28] At the heart of Warwick's thesis was a classification of nose shapes. Noses of the rich and famous were analysed according to a quasi-scientific assessment of their dominant and secondary shape. From him we learn that George Washington, the Duke of Wellington, and Rameses II had aquiline noses, revealing leadership, strength, moral courage, and honour. Straight noses were possessed by men of letters, such as Voltaire, Byron, and Shelley. Statesmen, artists (Michaelangelo), soldiers (Oliver Cromwell), scientists and theologians (Martin Luther)

[28] Warwick, E. 1848 *Nasology; or Hints Towards a Classification of Noses*. Richard Bentley, London. E-book: archive.org/details/nasologyorhintst00jabe.

sported strong, wide-nostrilled noses. The hawkish variety was apparently characteristic of people interested in the creation and distribution of wealth — the economist Adam Smith was a noted exemplar. Snub and upturned noses (sometimes called 'celestial' noses, because they point to the heavens) revealed different degrees of mental weakness in men, though Warwick grudgingly accepts them as attractive in women; James I and George I had snub noses, and James Boswell's upturned nose was no match for the strong, aquiline nose of Samuel Johnson LLD, his most illustrious subject.

Few women appear in Warwick's lists, though Queen Elizabeth I (aquiline) and the poet Felicia Hemans (straight) are notable exceptions. He notes helpfully:

> 'The energies and tastes of women are generally less intense than those of men; hence their characters appear less developed and exhibit greater uniformity. That their passions are stronger is undeniable, but these do not constitute character, nor are exhibited in the nose. Their indexes are the eyes and mouth, and therefore their consideration forms no part of the present subject.'

What utter nonsense! When Sultan Suleiman the Magnificent, ruler of the 16th century Ottoman Empire and the possessor of a fine hawkish nose himself, muttered in the ear of Roxelane, his wife: 'Is it possible that a little *retroussé* nose can reverse the laws of an Empire?' he conceded the influence she had over the affairs of State.[29] Her formidable character took her from the Sultan's bedroom to the position of chief political advisor to an Empire, without whose approval, no law was passed. The shape of her nose had nothing to do with it. Despite the awfulness of the 'science' behind nasology, and despite Mr Warwick's patronizing sexism and mind-blowing arrogance, he died a rich man; not the last charlatan to make a fortune from peddling fairytales passed off as science.

*   *   *

---

[29] Marmontel, J-F. 1791 *Contes Moraux, et Pieces Choisies* Vol 1. Paris E-book: archive.org/stream/contesmorauxetpi00marmuoft#page/n1/mode/2up.

Most animals don't have backbones and are classified as 'invertebrates', to distinguish them from 'vertebrates', which do. Over 99% of all animal species are invertebrates; only three-quarters of 1% are vertebrates. Although humans have few obvious affinities with invertebrates, the way our noses work, and the way their various smelling organs work, are really quite similar. Where they differ most obviously is in how they get odorant molecules to the olfactory membrane. Of the vast number of invertebrate species, only a few types have successfully colonized the land (and one extremely successfully), where smelling apparatus designed to catch airborne molecules is required. By contrast, with the exception of fish and some stages of amphibians' lives, all vertebrates breathe air, and so have smelling equipment that's stimulated only by airborne smells.

A significant evolutionary problem faced by all animals living on dry land, whether with a backbone or without, is the danger of desiccation, for life can't exist without water. Like insects, our waterproof skins stop the outward passage of water from our bodies, and we concentrate our liquid waste to ensure we lose as little water as possible. Our lungs, however, cannot function unless they are moist. The loss of water through respiration is a problem known to anyone who has travelled for several hours in a long-haul jet, where filters remove water vapour from the cabin air for reasons to do with aircraft safety, rather than passenger comfort. Our nasal membranes, too, must be moist for us to be able to perceive smell, and if they should start to dry out even a little, our sense of smell disappears. People visiting Antarctica, where the extreme dryness of the air removes water from the structures in the nasal passage as they breathe in, sometimes report a loss of smell. The effect is immediately reversed, however, when they're able to once again breathe humidified air.

If Noah had been given one grain of rice for every species packed into his ark, he would have been able to almost fill two 44-gallon oil drums. Our own species, *H. sapiens*, is just one of those 7.7 million kinds of animals he crammed into his hold;[30] indeed, all the animals with backbones together would hardly fill an ice-cream tub, removing

---

[30] Mora, C., Tittensor, D.P., Adl, S., *et al.* 2010 How many species are there on Earth and in the ocean? PLoS Biol 9(8): e1001127 doi:10.1371/journal/pbio.1001127.

but an insignificant amount of rice from his rice drums. Consider that every individual of every species employs its sense of smell every day for everything it needs to do. If we were to apply Socrates' dictum *Nosce te ipsum* — 'know thyself' — to the human sense of smell, we should first know how it's used by some of the vast numbers of animal species with which we share the Earth.

* * *

Animals detect smells in the world outside through a bewildering range of specialized chemical-detecting organs. Flaps and folds, hairs and pockets house chemosensory organs usually on, or around the front end of the animal, though this is by no means universal. How, and where, they're carried on the body depends upon the animal's lifestyle; whether a species lives in water or on dry land, whether it moves about or is sedentary, whether it pursues its prey or passively filters water, and so on. In aquatic animals smell organs are sometimes directly exposed to the water on fine filaments, while other times they are housed in pits or in other sheltered places. Marine snails bear them in shallow depressions under the awnings of their shells and directly in the pathway of water being pumped to the gills. Garden snails have them on the tips of their 'tentacles'; earthworms have them distributed along the length of their bodies, though densely arranged around their lips. Sea slugs carry them in ear-shaped flaps on their heads; crabs and lobsters carry their noses on little projections called 'antennules' sitting between their antennae, and octopus and their relatives have them around the suckers on their tentacles, and in small pits beneath their eyes. Figure 4.2 shows a few examples of invertebrate smell sensor equipment to give an idea of the range of structures that exist in nature.

* * *

The hold of Noah's ark would have been crammed with a dazzling array of species of the most successful group of animals ever to live on Earth — insects. There are estimated to be about 1 million kinds of insects, and at any one moment there are estimated to be 10 quintillion individual insects alive (10 followed by 18 zeros), outweighing all

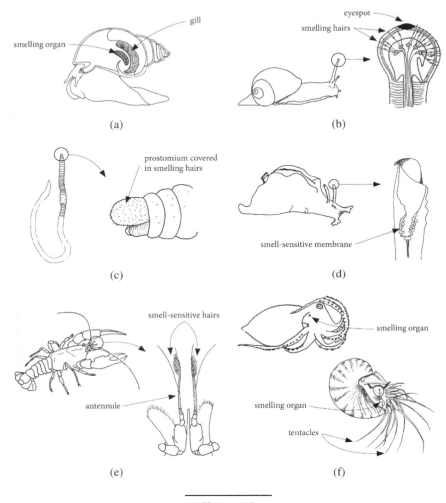

Figure 4.2

(a) Dog whelks, like all marine snails, carry their smell organs under the shell directly in the stream of water being pumped over the gill.

(b) In garden snails the smell organs are borne on the tips of the 'tentacles' There are tiny, hair-like projections covering the tip's surface. Each hair is connected to an olfactory nerve, *via* nerve ganglia, that runs down the 'tentacle' to the brain. Note the pigmented eyespot, right at the tip.

(c) Earthworms have smell sensitive hairs covering a fleshy lobe overlying the lips, called the 'prostomium', and the lips themselves are richly endowed with smell sensitive hairs. Smell signals are sent to the worm's brain, lying a few segments behind the mouth.

Fig 4.2 (Figure on facing page)

(d) Smell organs in the California sea hare are carried on two ear-like projections from the head, called 'rhinophores' — literally meaning 'nose carriers'. One of these is shown in detail. Smell-sensitive membrane lines the bottom of the pouch, from where an olfactory nerve projects to the brain *via* a number of nerve ganglia.

(e) Crayfish, in common with other crustacea, carry their smell organs on two small antennules, located on the head between the paired antennae. Tiny hairs on the fringed appendages capture smell molecules in the water as the fringe is extended and retracted.

(f) Smell organs in octopus, and their primitive relative *Nautilus*, are carried on the tentacles, and also in specialized olfactory pits lying below the eyes. (The sketch of *Nautilus* has been simplified by the omission of many tentacles.)

other land animals, including humankind. Insects have the most fantastic 'noses' in the entire animal kingdom, conducting nearly all of their life's activities through the sense of smell. They have been intensely studied in the laboratory and in the field, thanks to their economic impact as pests of crops and vectors of disease, and because of their value as pollinators. One famous species, the fruit or vinegar fly *Drosophila melanogaster*, was long ago adopted as a model species for the study of genetics. The genome of the fruit fly was sequenced in 2000, opening up great avenues for olfactory research in insects. The small size of insects makes them a little tricky to work with, but modern instruments and techniques for electrophysiological observations are overcoming this constraint.

Insects use smells to find their food from far away, to confirm food on contact, to lay trails for others to follow to exploit food, to advertise their sexual status, to locate a mate, to communicate alarm, to deter predators from attacking them, and to do other things as well. Not only do they react to environmental smells, like those coming from a food source, they also generate their own special scents, called 'pheromones', to coordinate reproduction. In the much-studied honeybee, a pheromone emitted by the single queen in the hive represses the development of the sexual equipment of other females, and if she's removed, one of the underlings quickly becomes sexually mature and takes over the task of egg production. Pheromones are unique to the species producing them, and necessitate smell receptors tuned to receive just them and not the pheromones of any other species.

Insects carry their smelling organs on their antennae, and occasionally elsewhere, such as on small tasting appendages, called 'palps', at the sides of the mouth. Antennae arise from the head and occur in a huge range of shapes and sizes, from feathery as in male moths, fan-shaped in cockchafer beetles, long and smooth in butterflies, beaded in longhorn beetles, to short and knob-like in grasshoppers (Figure 4.3).

Tiny hair-like structures on the antennae, called 'sensilla' (singular 'sensillum'), are responsible for detecting scent. Sensilla are the universal sensors for arthropods because they detect not only scents but also respond to vibration, sound, and even wind speed during flight. Irrespective of their use, all sensilla are built to a common plan, consisting of a hollow cone made of the same material as the insect's external skeleton. Their role is to extend the body's surface to where the particular sensation to be detected is encountered (Figure 4.4).

Once a smell molecule has entered the sensillum through one or more of its pores, it binds to a protein that carries it through the lymph surrounding the receptors to the projection of the olfactory cell. This happens very quickly; only between one and three thousandths of a second is needed for an odorant to reach the receptor from the outside of the sensillum. The odour receptors in insects are much like those in bacteria and humans, and do the same job of connecting the outside of the cell with the inside. The steps in the biochemical chain reaction between the receptor and the generation of a nerve impulse are a bit different from those in humans, resulting from the greater evolutionary age of insects than humans, (see Fig. I.1) but the outcome is the same. The sensillum is extremely sensitive; as little as a single molecule of female silkworm moth sex attractant is sufficient to trigger a nerve impulse in the antenna of a male. The number of sensilla is frequently not the same in male and female insects of the same species. In the flesh-fly, for example, females have five times as many sensilla as males, because it's the female that must find a decaying carcass upon which to lay her eggs. In the silkworm moth *Bombyx mori*, in which the male follows a plume of female mating pheromone, the male has 17,000 sensilla, while the female has only 6,000.

*   *   *

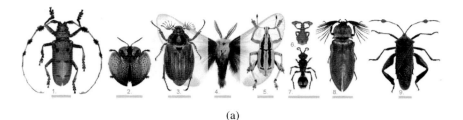

(a)

(b)

Figure 4.3

(a) The range of antennae types found in insects. (Scale bar =10 mm)
1. Longhorn beetle
2. Tortoise beetle
3. European cockchafer
4. Nevada buck moth
5. Snout beetle
6. Ground beetle
7. Velvet ant
8. Click beetle
9. Leaf-footed bug

Reprinted from Neuron, vol. 72, Hansson B.S., and Stensmyr M.C. Evolution of insect olfaction, pp 689-711, 2011, with permission from Elsevier.

(b) Male silkworm moth, *Bombyx mori*. Copyright Marcus Knaden, courtesy Max Planck Institute for Chemical Ecology, Jena.

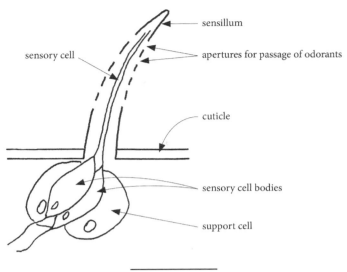

Figure 4.4

Insect sensillum. The waterproof cone has perforations large enough to admit scent molecules, but too small to allow the escape of water.

Stepping aside from insects for a moment, spiders and their eight-legged relatives carry their 'noses' on their legs. In the bizarre whip-scorpions — sometimes called whip-spiders — the first of the four pairs of legs is no longer used for walking but has become extremely long and thin. At their tips lie their smelling organs (Figure 4.5).

The 'whips' can be up to 300 mm long, giving this strange nocturnal arachnid a wide arc of surveillance. Common garden spiders have similar, though less exaggerated, smelling arrangements on their legs.

*   *   *

Smell is used widely by insects for finding food. Plants and animals can't help but leak smells from their internal and external surfaces, providing cues for insects on the lookout for food. The silkworm moth larvae, for instance, respond to a single odorant produced by leaves of the mulberry tree, called '*cis*-jasmone'. Their olfactory system has a finely tuned receptor protein for just this chemical. Cabbage white butterflies have receptors tuned to a group of compounds called

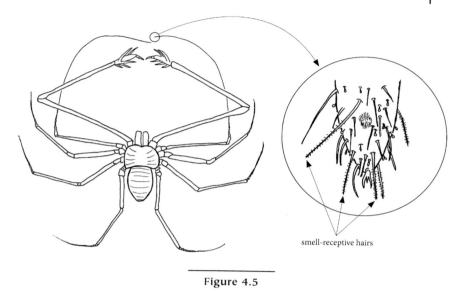

smell-receptive hairs

Figure 4.5

A tail-less whip scorpion, or whip spider, showing the greatly elongated first pair of walking legs. At their tips lie the smell receptive hairs, shown in detail below.

'isothiocyanates', by-products of the metabolism of members of the cabbage family. Insects that feed on a wide range of plants, like honeybees, must be able to learn a huge repertoire of flower scents and be able to add to it as new species become available. Not surprisingly, the part of the insect brain that is responsible for memory is larger in honeybees than in most other species.

Female mosquitoes, like many human parasites, have to feed on blood before they can lay their eggs. The *Anopheles* mosquito, which spreads the malaria parasite, is a significant scourge across much of the world, because human blood is favoured over the blood of other mammals. To find its prey, mosquitoes are finely tuned to perceive the smell of human sweat and have two proteins expressed in their olfactory tissue for this purpose. Only female mosquitoes feed and only females have the proteins. Intriguingly, mosquitoes infected with malaria parasites are significantly more attracted to human body smells than uninfected mosquitoes. Renate Smallegange and her colleagues have shown that under experimental conditions, infected mosquitoes settle much more frequently on nylon socks that have been worn for several hours than they do on unworn socks.[31] Uninfected mosquitoes show little difference in choice between worn or unworn socks. Just how the malaria parasite manipulates the mosquito's behaviour isn't known.

*   *   *

When it comes to sex and reproduction, insects show their supreme mastery of the world of scent. Just about everything from initial mate attraction to the protection of eggs is controlled by pheromones. The term 'pheromone' was first coined by biochemists Peter Karlson and Martin Lüscher almost 60 years ago working on butterflies. They defined the term 'pheromones' as: 'substances which are secreted to the outside by an individual and received by a second individual of the same species, in which they release a specific reaction, for example, a definite behaviour or a developmental process'.[32] The word comes from the Greek words *pherein* 'to transfer', and *hormōn* 'to excite', imparting the notion that pheromones act as external hormones. Karlson and Lüscher were careful to note that pheromones can be olfactory, when they operate over considerable distances such as in the sexual attraction of moths, or gustatory, when the transfer is effected orally, such as in the transfer of the so-called queen-substance in honeybees, which controls the production of various castes of bees in the hive. Insect pheromones are not only restricted to sex, but include chemical compounds produced for the expression of alarm, the establishment of a trail so others can follow, and for demarcating the boundaries of an individual's territory.

Insects, in common with almost all animals, are not continuously in reproductive mode. They mate at the time of year best suited to the development of their eggs and larvae. Some have very short periods when mating can occur, often characterized by vast swarms of

---

[31] Smallegange, R., van Gemert, G-J., van de Vegte-Bolmer, M., *et al.* 2013 Malaria infected mosquitoes express enhanced attraction to human odor. PLoS One 8(5): e63602.

[32] Karlson, P., and Lüscher, M. 1959 'Pheromones': a new term for a class of biologically active substances. Nature 183: 55–56.

males and females that may occur on just one, or perhaps on only a few days a year. This happens in mayflies, for example, in which all the adult males die over a period of a few days following feverish mating activity and the laying of eggs. Other insects, like butterflies and moths, live as adults for a few weeks during which time a mate must be found and the eggs laid in a place that will nourish the larvae. Whatever the life cycle, reproduction starts with the sexes coming together. The silk-worm moth shows how this happens.

*   *   *

The silkworm moth, *Bombyx mori,* has been bred in captivity for 5,000 years for the silk threads that wrap their cocoons. In 2007, the world's silk industry was worth over US $1billion, as reported by the UN's Food and Agricultural Organization, with China and India producing the lion's share. Intense breeding has yielded insects that produce much more silk than their ancestors and some strains are so large they can hardly fly. Like the majority of insects, it's the female moths that emit the attractive male-luring scent. The male silkworm moth's antennae are amongst the most elegant found in nature, curving outwards in two huge, graceful feathery arches (Figure 4.3).

Their vast surface area is required for the large number of sensilla with which they are equipped. Estimates in the literature of the distance over which a female can call a male vary from one to over 50 kilometres; what is important is not the quantum but that, given the right atmospheric conditions, a pheromone signal can be detected from a great distance.

The chemical composition of the silkworm moth pheromone was discovered by the German chemist, Adolph Butenandt, in the late 1950s and early 1960s. Twenty years earlier this remarkable scientist had shared the Nobel Prize in Chemistry for his work on human sex hormones. In 1961, Butenandt and his colleagues in Munich analysed and synthesized the silkworm moth's mating pheromone, naming it 'bombykol', after the moth's Latin name.[33] This was the first insect

---

[33] Butenandt, A., Beckmann, R., and Hecker, E. 1961 Über den Sexuallockstoff des Seidenspinners.1. Der biologische Test und die Isolierung des reinen Sexuallockstoffes Bombykol. Zeitschrift für Physiologische Chemie 324: 71–84.

pheromone ever chemically made in the laboratory. Bombykol is a volatile alcohol comprising 16 carbon atoms, and is produced in the paired pheromone glands in the abdomen of the female from hexadecanoic acid, a compound used by the moth for making the waxy material that keeps the cuticle waterproof. This happens through a simple chemical process, so the pheromone is biologically 'cheap' to produce.

Fifteen years later, a second sex attractant called 'bombykal' was discovered. Where bombyk*ol* is an alcohol, bombyk*al* is its related aldehyde. Aldehydes are produced by the oxidation of alcohols — thus the simplest aldehyde of all, formaldehyde, is derived from methyl alcohol — the simplest alcohol. Like bombykol, bombyk*al* is produced by the chemical reduction of hexadecanoic acid, and alcohol and aldehyde together constitute the moth's pheromone. The male moth needs to receive 200–300 molecules of the pheromone to show a behavioural response, though electrophysiological investigations in the laboratory show an antenna can respond to as little as a single molecule. When a female emits her sex-attractant pheromone, the gland is partially exposed and the pheromone is borne away on the wind. Silkworm moths, like most moths, call at dusk, when the light levels fall. When conditions for sex calling are right, a neuropeptide named 'pheromone biosynthesis activating neuropeptide', or PBAN, is released from the female's brain into its blood system, stimulating production of the pheromone. Experiments have shown that decapitated female moths do not produce the pheromone, but production can be induced if PBAN is injected into the abdomen of a decapitated moth. PBAN genes have been discovered in about 20 species of moths to control pheromone production, suggesting this control system might occur quite widely. Once mating has occurred, the female's brain shuts off production of PBAN, pheromone production is stopped, and she loses her attractiveness to males. In order to defend his future genetic investment, male moths usually deposit another odorous substance on the female that acts as a deterrent to keep other interested males at bay.

*   *   *

It's not always the females that initiate mating through the first release of mating pheromone. In a number of moths it's the males that

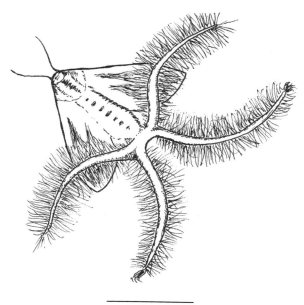

Figure 4.6

A male tiger moth *Creatonotus* sp., showing its scented coremata everted to attract a female. Many thousands of tiny hairs help the mating pheromone evaporate.

release the attractive pheromone. They do this from extraordinary structures at the tip of their abdomens called 'coremata'. Coremata are sacs attached to the inside of the abdomen that inflate when air is pumped into them, and can best be imagined as the inverted fingers of a rubber glove which pop out when air is blown into the cuff. They are covered in huge numbers of tiny hairs with an enormous surface area to facilitate the pheromone's rapid evaporation. The record for coremata size is held by the Asian tiger moth, in which they are one-and-a-half times longer than the male's body (Figure 4.6).

Pheromones dispersed by coremata-bearing species are synthesized from plant products the moths ate when they were larvae, and held in reserve for adulthood. Mostly the compounds are alkaloids — chemicals akin to those that give chilies their heat. Alkaloids are produced by plants as part of their chemical defences against being eaten; a defence easily overcome by specialist moths, it would seem, and put to good use. Caterpillars reared on high alkaloid diets have bigger coremata than those reared on alkaloid-free diets, and as larger quantities of pheromone are produced by larger coremata, they may serve as a signal to females of their genetic quality. Males of some coremata-bearing species aggregate at breeding time to combine their coremata scent displays in assemblages known as a 'leks', to which females from all around are attracted for mating.

* * *

Most vertebrate animals live on land, smelling the world as they take air into their lungs. They sport a huge range of nostril types and nasal cavities, specially adapted to their particular lifestyles. The most ancient vertebrates, however, are restricted to an aquatic life, smelling the outside world more in the way that invertebrates do. Fish detect chemicals with a structure called the 'olfactory rosette'. If you look at the snout region of a fish you'll see the entrance into the rosette chamber, just in front of the eyes. Depending on what type of fish you have, you'll see either a pair of small holes, or a pair of double holes. Occasionally, as in eels, there is a single water intake siphon separated by 10 mm from an outlet port, looking like a tubular nostril. Beneath the surface is a chamber filled with many thin plates, each of which is covered with tissue carrying olfactory receptors. Water forced into the inlet hole, or tubular nostril, as the fish swims then flows over the receptive membrane and is expelled from the outlet siphon. There are cilia in the rosette chamber whose beating helps the water flow through it, and some species of fish have the ability to lower the floor of the chamber a bit, so creating a sucking effect and helping the inflow of water at times when critical assessments are required. Water flows over the receptive membrane continuously and in this manner the rosette functions much as an insect's sensillum. It's never switched off and is constantly engaged in sensing the outside world. There are three types of olfactory receptors in fish rosettes, one of which responds to general odorants, such as the smell of the headwater stream in which the fish hatched, and food, and two that are specialized for detecting chemical cues associated with social and sexual biology (Figure 4.7).

* * *

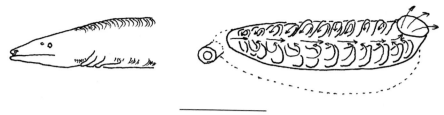

Figure 4.7

Head of an eel, showing inlet and outlet tubes to, and from, the olfactory rosette. At the right is shown a cutaway view of the course taken by water flowing between many thin plates lined with olfactory membrane.

Amphibians, reptiles, birds, and mammals live in, and on the surface of the Earth, and in the sky above it. Air enters their nasal passages *via* twin nostrils, which are sometimes highly specialized for protecting the nasal passages from water or sand, or for regulating the spot from where air is drawn, and sometimes for other things. On the whole, amphibians (frogs, salamanders, and a few other rather rare animals), reptiles, and birds have simply shaped nostrils, mostly being little more than orifices overlying the nasal passages. One or two desert dwelling reptiles have backwards-sweeping cowlings over their nostrils, to provide protection against blowing sand, but such adaptations are uncommon. Amphibians and reptiles have very active vomeronasal (VN) systems however, and in the most recently evolved reptiles — the snakes — the VN system dominates the sense of smell. The forked tongue — the iconic characteristic of snakes — picks up scent particles and transfers them directly into the paired VNOs, *via* two small slits in the roof of the mouth. Their nostrils do little more than allow the passage of air when their mouths are shut, though when a snake is struggling to swallow a large prey item, it is able to flip the end of its trachea out of the side of its mouth, bypassing the nostrils altogether while the back of its throat is blocked.

Birds are generally regarded as having a rather poor sense of smell, because mostly they are active during the day, and their frequently beautiful plumages suggests they live in a visual world. To an extent this is true; their eyes are finely tuned to discriminate favoured food items among a mass of inedible material, and those of them that predate on fast-moving and agile prey have excellent binocular vision. Their colourful plumage is used in sometimes very elaborate courtship rituals. The olfactory parts of the bird's brain are generally rather small and insignificant, and overshadowed by the parts that process visual input. Darla Zelenitsky and her colleagues have examined the evolution of the sense of smell in modern birds and compared them to their dinosaur ancestors, using a metric known as the 'olfactory ratio'.[34] This is the ratio of the length of a part of the brain known as the 'olfactory lobe' measured against the length of the whole brain. For species known only from fossils, the ratio can be determined by measuring the impressions of these structures in casts of their skulls. It has long been known that, at least for living birds whose behaviour can be studied, larger olfactory ratios are positively correlated with a better ability to smell. What emerges is that the olfactory ability of ancestral birds at the end of the age of the dinosaurs rose, and maintained a plateau of effectiveness through the mass extinctions at the end of the Cretaceous epoch, about 65 million years ago. This enhanced sensory ability might have contributed to the survival of the reptilian-like ancestral birds (*Archaeopteryx, Ichthyornis* etc.) that emerged after the dinosaurs had disappeared.

As modern birds evolved, olfactory ratios began to decline, and in the most recently evolved birds, such as the songbirds, parrots, seabirds, eagles, and owls, they suggest a decline in importance of the sense of smell. In the most ancient species of living bird, the New Zealand kiwi, the sense of smell is still very good. The kiwi uses its nose to find invertebrates hidden in the soil and is, incidentally, the only bird to have its nostrils at the very tip of its beak (Figure 4.8).

As in most biological systems there are some outstanding exceptions to the general rule, and one of them is the North American turkey vulture. This bird is adapted to locate dead animal carcasses from great heights. To do this, its nose picks up odours of putrefaction in upwelling air, inviting it to descend close to the carcass and feed.

---

[34] Zelenitsky, D.K., Therrein, F., Ridgely, R.C., *et al.* 2011 Evolution of olfaction in non-avian theropod dinosaurs and birds. Proceedings of the Royal Society B 278: 3625–3634.

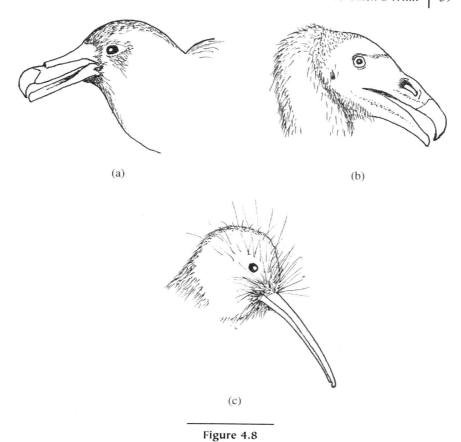

(a)

(b)

(c)

Figure 4.8

Heads of the (a) giant petrel, (b) turkey vulture, and (c) kiwi, showing nostril type. The giant petrel's nostrils lie in twin tubes lying along the top of the bill. Turkey vultures have wide nostrils that penetrate the bill from one side to the other, to assist them clearing blood and tissue from their noses as they feed. Kiwis are the only bird species to have its nostrils located at the very tip of its bill, enabling it to smell out invertebrates deep in the soil.

Their olfactory ability is so good that when the Union Oil Company of California tested the welds of its oil pipeline in the 1930s, it did so by introducing sulphurous-smelling ethyl mercaptan — a smell of decaying flesh — into the flow line. Leaky welds were quickly sniffed by turkey vultures which flew down and sat on the pipeline right where the leak was detected. Patrolling engineers could quickly identify the trouble spots and fix them (Figure 4.8).

The genetic analyses of olfactory receptors in birds' noses show that gene repertoires are quite large and the proportion of pseudo-genes is low, suggesting that olfactory ability is far from negligible. Night active species, such as the kiwi and the New Zealand flightless kakapo, have larger gene repertoires than other birds, consistent with the idea that they find food with their noses. Yet birds with quite small olfactory receptor gene repertoires, such as the common starling, are still able to detect and discriminate volatile compounds in the leaves of food plants. The final stages of navigation by homing pigeons are guided by smell, although long-distance navigation depends upon their ability to orientate according to the position of the sun and stars. Seabirds in the family known as the tube-nosed birds (fulmars, pet-rels, shearwaters, and albatrosses), on account of a tubular manifold surrounding the nostrils, are the super-smellers of the bird world (Figure 4.8). Their tubular noses increase the efficiency of the olfac-tory membrane by pressurizing the incoming air, making it easier to locate food in the ocean. They respond to low concentrations of fish and krill oil emanating from the sea's surface, doing this in gale force winds when ships struggle to maintain course. When they return to land, they locate their individual burrows by their unique smells. They appear to have well-developed odour memories, because they are able to return to their old breeding burrow, year after year.

*   *   *

The ways in which dogs take air into their noses has been extensively studied because of the incredible olfactory abilities of canines. Dogs are said to be 'macrosmatic' — that is, they have a very sensitive sense of smell and are able to detect scents at extremely low concentrations. Species with a poor sense of smell are said to be 'microsmatic'; old texts tend to place humans in this category. These terms are too sim-plistic, since many species with anatomically simple noses exhibit extreme sensitivity to certain odorants. Because the terms are in com-mon biological use I'll use them, but bear the above caveat in mind.

For humans to be able to smell a particular odorant, there has to be lots of it available. Dogs, on the other hand, can detect odorant molecules at a 10,000 to 100,000 times lower concentration than we can. This is due to four factors. Firstly, dogs have between 125 million and 300 million olfactory receptor cells; humans have 5–6 million. Big dogs, like bloodhounds, are at the upper end of the range, with small dogs, like dachshunds, at the lower. Secondly, the area of the nasal cavities covered by the olfactory membrane measures about 200 cm^2 in dogs, compared with only 6.5 cm^2 in humans — this can be imagined as a postcard compared to a postage stamp. Thirdly, dogs have much better developed nasal cavities than humans, in which incoming air is streamed either to go down into the lungs, or to be subject to smell analysis. This system allows the dog to breathe quickly, as it would need to when following a trail while constantly analysing scent. In humans there's no such separation of air streams; if we breathe fast our olfactory membrane has little time to analyse each new breath. And fourthly, as we saw earlier, dogs have about twice the number of functional olfactory genes than we do.

The olfactory membrane of all mammals is subject to overstimulation, a phenomenon known as 'olfactory fatigue'. Your nose becomes fatigued when you go into a perfume shop, for example, where the air is laden with many fragrances and you are very aware of the ubiquitous scent. In a few moments, however, you cease to notice the background smell and can concentrate on choosing a perfume. For macrosmatic animals olfactory fatigue is an important matter; their noses must be on duty all the time if the information they are supplying to the brain is to be trusted. Mammals overcome this problem by sniffing — taking repeated short bursts of air with brief periods of calm in between, sufficient to allow the membrane to clear itself of what was there before. When dogs first detect an interesting scent they change their respiration rate to short, sharp sniffs, lasting about a half second. Later they may lengthen their inspirations to a second or more. Their rate of sharp, short sniffs is an order of magnitude faster than in humans, and they handle much more air per unit time than we do.

The nasal cavity of the dog, and of all macrosmatic mammals, is very different from that of humans. The most obvious characteristic of the faces of macrosmats is that they have long, broad snouts, while humans and apes are characteristically flat-faced. The dog's snout is divided into three chambers, one right at the front where incoming air is received, one that conditions, humidifies and warms the air, removes foreign bodies, and generally cleans it before it continues on its way to the lungs, and a final one (called the 'nasal recess') that is dedicated to smell detection (Figure 4.9).

In humans, the task of cleaning the incoming air is undertaken by short, stiff filtering hairs growing just inside the nostril. Next, the air is passed over flaps of respiratory membrane, to warm and humidify it. The respiratory membrane also kills off bacteria and other organisms that could damage the delicate lung tissue. Finally, as the inspired air is drawn down the back of the throat (called the nasopharynx) and into the lungs, some is turbulently thrown up to the ceiling of the back part of the nasal cavity, where it passes over the olfactory membrane. Only about 10% of inspired air is sent to the olfactory membrane, leaving 90% unsampled. Not much of the sampled air will continue down into the lungs, instead it will be joined by expired air on the next out-breath, and expelled *via* the nostrils.

There's some compensation for the simple structure of the human nose, however. Because we lack the equipment to stream the incoming air, we're capable of 'retro-nasal' smelling; that is, we are capable of passing air from the mouth to the nasal cavity *via* the back of the pharynx. Much of our enjoyment of food comes from its smell, detected as we chew, just as much as from its taste. We'll see later that retro-nasal smelling has played a big part in the facilitation of human dining culture, something that dogs couldn't do. Every dog owner will attest there is little evidence a dog vacuuming up its food savours it at all.

Brent Craven and his colleagues at Penn State University have used the technique of functional magnetic resonance imaging (fMRI) to investigate the structure of a dog's nasal cavity. They have combined their anatomical findings with the application of how fluids behave, and built a model of how a dog's nose works.[35] Air entering

---

[35] Craven, B.A., Paterson, E.G., and Settles, G.S. 2009 The fluid dynamics of canine olfaction: unique nasal airflow patterns as an explanation of marcrosmia. Journal of the Royal Society Interface 7: 933–943.

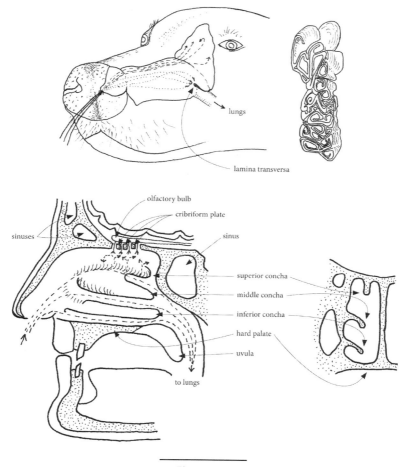

Figure 4.9

Three-dimensional model of the left nasal airway of a dog, built from high-resolution MRI scans. Immediately inside the nostril lies the nasal chamber, where air is injected into the respiratory chamber. This chamber is lined with respiratory membrane to condition the incoming air. At the rear of the snout the olfactory membrane covers a honeycomb-like structure of thin bones (called the 'ethmoturbinals'). This is shown on the right. Air sent for sniffing (dashed line) travels along the upper part of the respiratory membrane, up to four times faster than air sent to the lungs (dotted line). Redrawn from Craven B., *et al*, 2005, Journal of the Royal Society Interface vol 7 doi: 10.1098/rsif.2009.0409

Compare the nasal passage of the dog with that of a human, in which the nasal bones are simple in structure, and with little difference in air speed between air sampled by the olfactory membrane and that sent to the lungs. In humans, olfactory membrane covers only the superior concha (ethmoturbinal).

the nostril is immediately organized into two streams; one destined for the lungs and one for the nasal recess. The recess stream, which keeps close to the roof of the nasal cavity, moves more quickly than the respiratory air. In the nasal recess olfactory membrane lies over a fine reticulation of filigree-like bone of immense area, packed into a tiny volume (Fig. 4.9, top). The slower moving respiratory stream travels centrally through the snout, enters the nasopharynx and travels onwards to the lungs. A small flap of bone lying at the base of the nasal cavity, called the *lamina transversa*, deflects expired air forwards into the central part of the snout as it leaves the nasopharynx, leaving the air that was sent for odour processing in relative calm, to be expired only when fully analysed. Humans lack this sliver of bone, with the consequence that some expired air always travels upwards to the olfactory membrane before being expelled.

*   *   *

The respiratory membrane that cleans and conditions incoming air is a delicate tissue, which must periodically rehydrate itself and dispose of waste products extracted from the air flow. It must also allow replacement of its stocks of antibiotics used to attack organisms that might damage the nasopharynx and lungs. To enable this to happen the two sides of the nasal cavity undergo what's called a 'nasal cycle', in which one half becomes congested, slowing or even stopping the airflow, while the other remains open for business, before shutting the open side and opening the congested side. The cycle is driven in part by a structure at the anterior of the cavity called the 'nasal swell body', as well as by a cyclical thickening and shrinking of the respiratory membrane itself. Figure 4.10 shows a section through the human nasal cavity, in which the right side of the cavity is congested while the left side is open. In humans the nasal cycle takes between 50 minutes and 4 hours to complete, with an average of about 2.5 hours, and may help explain why you turn over regularly during sleep. The length of the cycle in dogs isn't known, but in rats and rabbits it's between 30 and 85 minutes in the former, and between 80 and 150 minutes in the latter.

*   *   *

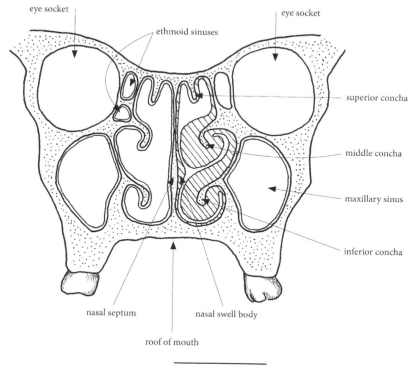

Figure 4.10

Vertical section through the human nasal cavity, at the level of the eyes. The left side allows free passage of air, while the right side is congested. The respiratory membrane on the right side is shown cross-hatched.

You'll be very familiar with the shape and structure of your own nostrils, but the chances are your knowledge of the nostrils of other animals is a bit sparse. I might be safe in assuming you haven't examined the nostrils of a tube-nosed bat, or a star-nosed mole, to mention but two of many species with unusual nostrils. The nostrils of mammals are works of art, belying their important functional role in smelling. Most mammals have nostrils not unlike the familiar soft, moist noses of dogs and cats. Far from being simple holes, however, their nostrils have sculptured side flaps that move and vibrate as the animal sniffs the ground, and internal folds that can be raised and lowered during the breathing cycle. Mammals from the high arctic and tundra have dry, fur-covered nostrils to prevent them from

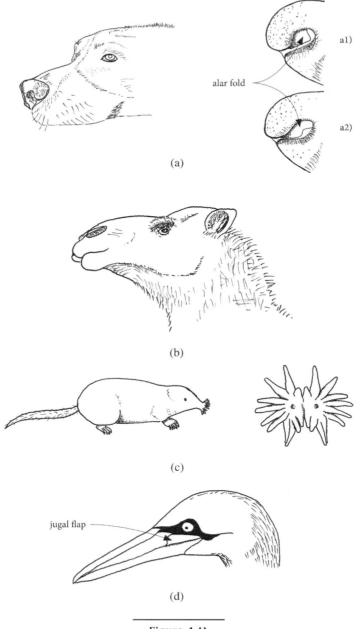

Figure 4.11

Figure 4.11 (Figure on facing page)

(a) The dog has a hairless, usually damp, nose pad (called a rhinarium'), perforated by two comma-shaped nostrils. In the detailed sketch, a1) shows the alar fold's position during sniffing, a2) during exhalation.

(b) Camels, like other desert species, have hairy nostrils that can be closed to prevent sand entering the nasal passages.

(c) The star-nosed mole of eastern USA has simple nostrils surrounded by fleshy lobes, which can detect the slightest vibration made by a worm or invertebrate.

(d) The gannet has no nostrils at all, instead taking in air through a flap (called the 'jugal flap') at the side of the bill. This moveable plate can be closed tightly, allowing the gannet to dive from 60 m or more without inundation.

freezing, as can be seen in horses, camels, and llamas. Desert species, such as African camels, can close their nostrils to keep out blowing sand. Swimming mammals, such as moose, beavers, and capybaras from the Amazon, are able to close theirs when negotiating rivers and swamps. Some mammals have nostrils modified for use as food-gathering devices; the elephants' trunk is a well-known example; the star-nosed mole's finger-fringed snout, used to locate soil-dwelling invertebrates, is a less well-known one (Figure 4.11).

Nostrils come into their own when air is expired. Gary Settles and others at Penn State University have examined how they work, using a variety of flow visualization techniques.[36] Taking the dog as an example, the shape of the nostrils isn't the same during breathing in as when the dog breathes out. During inspiration, a small membrane inside the nostril, called the 'alar fold', and just visible at the upper edge of the nostril, is drawn back, (Figure. 4.11 a1) allowing air to enter the nostril near the top of the orifice. As the dog breathes out (Figure 4.11 a2) the membrane descends, deflecting the expired air laterally and downwards, crucially directing it away from the odour source, and thus not disrupting the aerodynamics of sniffing. By directing the air stream away from the scent, fresh air from the odour source is drawn towards the nostril in response to the slight

---

[36] Settles, G.S., Kester, D.A., and Dodson-Dreibelbis, L.J. 2002 The external aerodynamics of canine olfaction. In: Barth, F.G., Humphrey, J.A.C., and Secomb, T.W. *Sensors and Sensing in Biology and Engineering.* Springer, New York.

lowering of pressure created by the jet of expired air, giving the nostril a virtual forward reach of several cms (Figure 4.12).

The way the alar fold directs the exhaled airflow ensures that inspired air isn't contaminated by expired air. Such variable-geometry nostrils are common in mammals and are a regular feature of macrosmatic species. Our nostrils have no trace of an alar fold and as a result we have no control over where our expired air stream goes. Breathe out through your nostrils on a frosty morning and try to direct the stream of steam by wriggling your nose to see what I mean.

While it's true that the primary function of the nostrils is to enable air to be drawn into the lungs *via* the nasal chamber, the fact that they are paired, and that the nasal cavity is separated into two distinct halves, suggests they may play a role in separating left and right sensations, just as the eyes and ears perceive light and sound stereoscopically and stereoacoustically. Stereo perception allows the stimulation from two slightly separate locations to be compared by the brain, enabling the stimulus' source to be accurately pinpointed. Of course, a flexible neck like ours allows the head to be swung from left to right, and the nostrils to draw air from different locations, but not all vertebrates have necks as flexible as ours. If fish were to compare odours from left and right, for example, they would have to move their whole bodies from side to side.

Consider what you do when you are choosing a perfume in a department store. You are handed a paper smell strip by the assistant and you start your work. You slowly move it from left to right underneath your nostrils, while taking tiny sniffs. You could — but most probably wouldn't — hold the smell strip steady and wag your head from side to side, like a dog chasing a scent (if you opt for this method I'd recommend you find a spot out of the view of other shoppers!). By wafting the scent strip from left to right, you mimic the effect of wagging your head from side to side, allowing sensations from left and right to be compared. You'll notice you subconsciously expose all interesting smells, and not only perfume strips, alternately to your left and right nostrils as you investigate a smell.

Evolution in some species has resulted in a widening of the base between their smelling organs, enabling them to accentuate the difference between left and right input. Hammerhead sharks have,

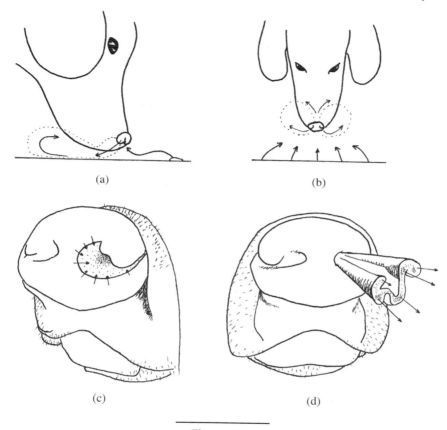

(a)   (b)

(c)   (d)

Figure 4.12

(a) and (b) As a dog exhales, air is directed backwards and upwards, away from the
source. This has the effect of decreasing the air pressure immediately in
front of the nose, so drawing unsampled air towards the nostrils. The
dynamics of inhaling and exhaling are shown in (c) and (d), respectively.
During inhalation, air travels slowly from a wide sampling arc, gaining
speed as it approaches the alar fold. Air is exhaled in a trumpet-shaped
flare in which some parts move faster than others. Length of arrows, and
density of stippling denote air speed.

(a) and (b) are redrawn from Settles G.S., *et al.* 2002 in *Sensors and Sensing in Biology
and Engineering*, Springer, New York, (c) and (d) are modified from Craven B. *et al*,
2005, Journal of the Royal Society Interface vol 7 doi: 10.1098/rsif.2009.0409

arguably, the most widely spaced nostrils — relative to head width and overall body size — of all vertebrate animals. Their nostrils are borne at the ends of wide lateral projections of the head, called 'cephalofoils', separating them by a metre or so (Figure 4.13).

Hammerheads have olfactory rosettes, like those of other fish, located close to the ends of their cephalofoils, positioned right behind the foil's leading edge. Water is fed into each rosette *via* a groove that collects the fast-moving water created by the front of the cephalofoil and channels it into the rosette. There is little doubt that widely spaced nostrils provide better spatial sampling than is the case in other sharks.

The little tube-nosed bat weighing only 30–40 g has quite a flexible neck, but wagging the head from side to side while flying creates problems with aerodynamics and directional stability. The nostrils are extended laterally into tubes that sample air from locations about 10 mm apart which is a wider separation than would occur if they were positioned close to the skull. This arrangement allows tube-nosed bats to identify a single ripe fruit in a hand of bananas, and accurately locate where a meal is to be found. A time difference of as little as 50 milliseconds (50 thousandths of a second) in the arrival of odorants at the left and right nostrils is sufficient to enable macrosmatic mammals to pinpoint the source of a smell.

*   *   *

This is as good a place as any to address an often-asked question, and that is 'How good is the human sense of smell?' For the past several decades the perceived wisdom was that humans could detect 10,000 different smells; a figure that was based on woefully inadequate understanding of the sense of smell, and enthusiastically taken up by the media. Recent research shows that humans can discriminate at least one trillion different smells, and possibly many more.[37] This gives the sense of smell considerably more discriminatory power

---

[37] Bushdid, C., Magnasco, M.O., Vosshal, L.B., *et al.* 2014 Humans can discriminate more than 1 trillion olfactory stimuli. Science 343(6177): 1370–1372.

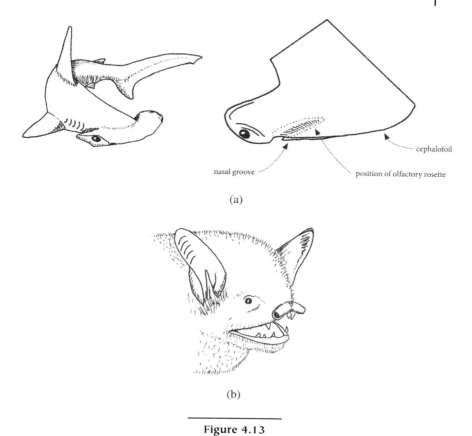

nasal groove

cephalofoil

position of olfactory rosette

(a)

(b)

Figure 4.13

(a) Hammerhead sharks carry their nostrils at the ends of a flattened head plate (called a cephalofoil'), which adds width between the nostrils.
In the detailed sketch, an olfactory rosette lies inside the cephalofoil, and receives water directed into it by the nasal groove.
(b) Tube-nosed micro-bats have nostrils are dawn out into two moveable tubes, enabling air to be sampled from a wide base.

than sight or hearing, turning the old view on its head. The real answer to the question, of course, depends upon what you want to use it for, or rather, for what purposes it has evolved. If the evolutionary requirement is for a nose capable of detecting tiny amounts of certain smells among many others, for example, we fail miserably, outclassed by even the smallest dog. If the requirement is for a nose able to discriminate between many different smells, however, we perform excellently. Gordon Shepherd at Yale University argues that the human sense of smell works better than it should, in terms of its reduced number of functional olfactory receptor genes, small nasal cavities, and a less-than-perfect system for getting scent-laden air to the olfactory membrane because of the number of central brain structures involved in odour processing.[38] Much of the most ancient part of our brains (in evolutionary terms) is employed to this end. Shepherd notes that the most complex olfactory tasks involve the temporal and frontal lobes of the brain as well as the ancient, reptilian brain, parts wherein lie higher cognitive abilities, including odour memory. The result is that humans, alone of all mammals, can pull more cognitive power into odour analysis and processing than any other mammal, with the astounding consequences noted above.

There are a few published anecdotes of where some central brain disturbance results in enhanced olfactory perception. One of the best known is by the neurologist Oliver Sacks in his essay titled *The Dog Beneath the Skin*.[39] He reports the case of Stephen D, a 22-year-old medical student, high on cocaine and angel dust. After a night spent dreaming he was a dog, Stephen D woke to find that his sense of smell was now like that of a dog. He described his scented world 'as if I had been totally colour-blind before, and suddenly found myself in a world full of colour'. His world was transformed. While the psychedelic drugs had enhanced all his senses, it was the change to his sense of smell that had the biggest impact. Sacks quotes him as saying: 'And with all this there was a sort of trembling, eager emotion, and a strange nostalgia, as of a lost world, half forgotten, half recalled.' Stephen D could recognize people by their scent, just like a dog, and every street and shop in his neighbourhood was recognizable by its smell. He had an urgent craving to smell everything he touched, as if

---

[38] Shepherd, G.M. 2004 The human sense of smell: are we better than we think? PLoS Biology 2(5): 572–575.
[39] Sacks, O. 1985 *The Man who Mistook his Wife for a Hat and Other Clinical Tales*. Simon and Schuster, New York.

a full understanding of it demanded its smell be examined. After three weeks he abruptly returned to his pre-trip state, and his perception of the world returned to how it had been before. He said: 'I'm glad to be back, but it's a tremendous loss, too. I see now what we have given up being civilized and human. We need the other — the "primitive" — as well.' Sacks tells us that the condition of his patient is not unlike what happens during an 'uncinate seizure', a type of temporal lobe epilepsy. What this anecdote shows is that the human nose, with all it anatomical limitations, is nevertheless capable of providing sufficient information for the brain to paint a much enhanced odorous picture of the world than it does in everyday life, when uninfluenced by mind-altering drugs.

While Sacks reported his subject experienced a vibrant and exciting new sensory world while under the influence of drugs, people who experience a loss of their sense of smell through injury or disease report a lack-lustre world, stripped of texture and interest. Many patients who have lost their sense of smell following a knock on the head (whiplash accidents were a common source of such losses before car headrests became commonplace) go on to recover some, if not all of their sense of smell, and frequently report that only when they were without their noses did they realize how much the world smelled, and how important this background smell was to their normal functioning and feelings of well-being.

Putting all the scraps of evidence together we can conclude that each olfactory receptor unit in humans is much the same as each receptor in dogs or mice. We have fewer of them, and fewer types of them (see Fig 3.2), and the area of our olfactory membrane is very small. Our nasal cavities are of simple construction, lacking the sophistication of those of dogs and other macrosmatic species, but we make up for these deficits by devoting much more of our brains to analysing and amplifying the signals. In terms of sensitivity to certain types of chemicals, human noses are as good as those of rats and dogs. Like fruit bats and fruit-eating monkeys, our sensitivity to fruit odours is high, recalling our distant evolutionary past when our ancestors called the forests home. We'd have to conclude that humans have a good sense of smell, at least for a limited range of odorants.

Although humans have only one of the two parts of the mammalian sense of smell, the brain has evolved to give meaning to information received from both. In the next chapter we'll look at the smelling brain, and start to understand how the nose and VNO separately stimulate the emotional brain, and what humans have lost — and gained — by not having a functional VNO.

Chapter **5**

Making Sense of a Sense

While it's the sense organs that gather information about the outside world, it's the brain that turns their signals into meaning. This is as true for silkworm moths, with their tiny brains, as it is for you and me, with our massive brains. All mammals have large brains and, relative to body size, ours are about as big as they come. Computed tomography (CT) scans of fossil mammal-like reptiles, dating back 190 million years ago, reveals brains that were relatively half as big again as those of the reptiles with which they co-existed and, interestingly, the increase occurred to the part of the brain that handles smells. CT analysis of the earliest true mammal known, a shrew-sized creature called *Hadrocodium*, revealed a relatively enormous olfactory brain with wide nerve tracts leading from it to the rest of the brain. Why would a large brain, and particularly a large smelling brain, be important to mammals?

More than other vertebrates, mammals are essentially social animals, spending much time communicating with their peers. Even the so-called 'solitary' species, i.e. those species that live most of their lives on their own, such as foxes, leopards and rhinoceros, must interact and communicate with others for finding a mate, and performing elaborate and sometimes protracted courtship rituals. In common with every single one of the 5,400 mammalian species, the newborns of solitary species must interact with their mothers, upon whom they depend for food until they are mature enough to fend for themselves. Communication between individuals, upon which social behaviour is built, was originally driven by smells, for early mammals were active only at night. They shared the Earth with a wide range of predatory reptiles that slowed up at night when the temperature dropped, leaving the stage clear for the newly emerging, warm-blooded mammals to make a living by moonlight. The growth of the mammalian brain allowed more complex smell communication to take place, as the social lives of early mammals became ever more complex.

Social interactions are controlled by the emotional part of the brain, which also happens to be where the sense of smell sends its signals, so the processing of smell signals and emotions have shared a long evolutionary history. Bettina Pause at the University of Düsseldorf argues that the role of smell communication in human evolution may well have been underestimated, because in modern humans it's seldom conscious. She emphasizes 'that in order to adjust to the environmental social requirements, social signals commonly require a fast and automatic response and [need to be] processed implicitly without the allocation of [extra, brain] resources'.[40] The primates continued along the social communication path with increasingly complex social behaviour requiring increased emotional processing power. The result, heavily elaborated in modern humans, is an enormous brain.

*　　*　　*

The mammalian brain handles smell inputs from two sources — one from the nose, and the other from the VN system. It's in the brain that nervous stimulation from each system is organized into information, which is then passed to the appropriate implementation centre. If the smell is a warning signal indicating some food is bad to eat, the information passes to the centre responsible for feeding behaviour, and the food is rejected. If it's a sex smell, the information passes to the centres of the brain where sexual behaviour is released, and so on. The nose and the VNO have quite separate, distinct pathways into the brain, utilizing different nerves and targeting different locations

---

[40] Pause, B.M. 2012 Processing of body odor signals by the human brain. Chemical Perception 5: 55–63.

in the brain. Often the nose and VNO project to the same centres, but generally they target different parts of those centres.

The pathway from the nose to the brain is anatomically simple. When you smell something, molecules of the odorant bind to the olfactory receptors embedded in the olfactory membrane in the nose. Nerve impulses generated by the receptors pass along the olfactory nerve, entering the skull through a wafer-thin, sieve-like bony plate at the front of the brain, lying underneath the inner corners of the eyes. Right behind this so-called 'cribriform plate', named from the Latin 'cribrum' meaning sieve, lie a pair of olfactory bulbs — the first way-station on an impulse's journey to its destination, and the first point in the brain to receive olfactory input. In fish and reptiles the olfactory bulbs are quite large, dominating the front part of the brain, and representing almost 20% of the brain's total mass. In humans the olfactory bulbs are tiny, hidden away under the frontal part of the brain. Despite their small size, they perform the vital editorial function of processing smell signals into meaning.

Each olfactory bulb houses about 2,000 bundles of cells, called 'glomeruli'. The five million receptors in the nose send their signals to glomeruli that are tuned to receive the type of signal to which the receptors respond. After activation, glomeruli communicate with others in their neighbourhood responding to the same odorant. The result is a pattern of glomerulus firing that creates what can be regarded as an odour map, or an organization of bits of information into meaning. Think of the olfactory bulb as a grid of light bulbs with each glomerulus being represented by one light bulb. Different smells result in different patterns being lit up. If the grid holds 2000 light bulbs, the number of patterns that can be created is almost limitless. The exact details of olfactory processing are not fully understood, but it's in the olfactory bulbs that it happens.

After the processed signals leave the olfactory bulb, they travel to the central part of the brain known as the 'limbic system'. The limbic system gets its name from the Latin 'limbus' meaning 'border', for the limbic system forms the inner border to the brain's central, basal part. If you imagine a line passing through the head from one temple to the other, and another at right angles to it from the bridge of the nose to back of the head, their intersection would be right in the middle of the limbic system. The system is a collection of separate parts that work as a whole, and represents the most ancient vertebrate brain. It's sometimes called the 'reptilian brain', because of its evolutionary antiquity. The brains of fish consist of olfactory bulbs, olfactory processing structures, and little else. As vertebrates evolved, their brains became more complex with other bits and pieces added until, as in humans, the limbic system is physically overhung by the great mass of the cerebral hemispheres that fills most of the skull.

The limbic system is primarily responsible for emotions and instinctive reactions, and for the formation of memories. It comprises many individual structures, each with their own special tasks to perform, but all parts work in concert with one another. They are linked together such that a constant 'chatter' is maintained around the system. Much of what goes on in the limbic system still remains a bit of a mystery.

Nerve impulses from the nose project into four main centres in the limbic system, and each of these connects with additional structures within, and outside, the limbic system. These are shown schematically in (Figure 5.1).

After leaving the olfactory bulb, nerve cells project into the 'olfactory tubercle' and the 'piriform cortex' (sometimes called the 'olfactory cortex' or 'olfactory lobe'). Exactly what the olfactory tubercle does is unclear, but it's involved in arousing the animal to incoming signals needing attention. It's also involved in the reward system of the brain and may be activated in addictive situations. The piriform cortex is where much of the fine processing of odour signals takes place. The piriform cortex sends messages to several places, including to the 'hypothalamus', a part of the brain that plays a central role in sexual physiology and behaviour.

Another of the piriform cortex's onward projections is to a structure lying outside the limbic system called the 'thalamus', a part of the brain that can be likened to a telephone exchange from where messages are sent out from the emotional to the rational brain. If you smell a flower and want to recall its name, for example, messages about the smell go from the piriform cortex to the thalamus, and

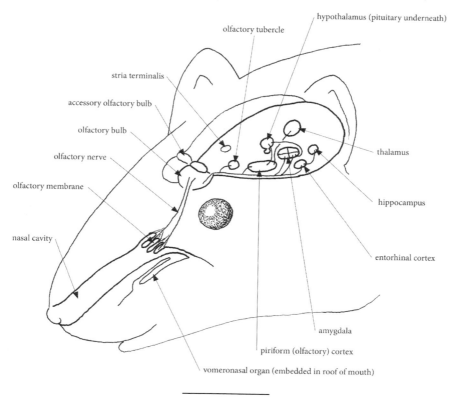

hypothalamus (pituitary underneath)

olfactory tubercle

stria terminalis

accessory olfactory bulb

olfactory bulb

olfactory nerve

olfactory membrane

nasal cavity

thalamus

hippocampus

entorhinal cortex

amygdala

piriform (olfactory) cortex

vomeronasal organ (embedded in roof of mouth)

Figure 5.1

Schematic showing the main olfactory nerve pathway to the key parts of the smelling brain in mammals. For clarity, the pathway is shown only on side of the brain.

onwards from there to a part of the brain called the 'orbito-frontal complex', located behind the forehead. The orbito-frontal complex of the brain is involved in what has been termed 'executive functioning' of the body, responsible for all the smell-related activities associated with determining if a smell is familiar or novel, and if a particular food is edible or should be rejected. It's the part of the brain where smells are rationalized, where volcanic emotional outbursts are restrained, and where everything smelled is subjected to forensic scrutiny. It'll prod your memory to let you remember the name of the flower. The orbito-frontal cortex is also responsible for impulse control, encouraging you to act in accord with social norms, enabling you to plan future responses and actions before you take them. In other words, it stops you from shooting first and asking questions afterwards.

The other two centres to which the smell neurons from the olfactory bulb project into are structures called the 'amygdala' and the 'entorhinal cortex'. The amygdala, (from the Latin for 'almond', reflecting the structure's shape) is a complex structure with at least four main parts and the nose projects onto two of these. The amygdala is involved, among many things, in emotional responses, particularly fear, and aggression. It's also involved in the processing of social information and in sexual arousal. Lastly, olfactory nerve cells project to the entorhinal cortex. This structure forms a bridge between what the nose smells and what the brain remembers about them. It sends nerve cells to a part of the brain called the 'hippocampus' (Greek for 'sea horse' — again, in recognition of the structure's shape), which is heavily involved in the storage and recall of memories, including memories of smell. These, then, are the principle targets for information coming from the nose. The contacts each part makes with other areas of the brain, often including its partner on the other side, complete the complexity of the smelling brain.

Astonishingly, there are just two nerve cells spanning the distance between the nasal cavity, where the scent molecules encounter the olfactory receptors, and the limbic system inside the brain. In the eye and ear, there are multiple cells in the nerve chains running from the eye's retina, and the ear's cochlear organ, to the brain. The neural simplicity in the sense of smell is a measure of its ancient origin, with its basic wiring retaining its original structure despite enormous evolutionary changes affecting the bodies of animals.

There's another striking difference between the nose, and the eye and ear. Scent signals pass directly from the nose to targets inside the emotional brain, without stopping off anywhere else before they reach it. In the eye, ear, and other senses, signals pass first to the thalamus (the 'telephone exchange'), and then to specialized processing centres elsewhere in the brain, such as the visual and auditory centres, as well as the limbic system. Significantly, the thalamus receives *no direct input* from the nose, only receiving referred signals from the piriform cortex. This means that your first response to a smell is always emotional; only secondarily do you rationalize it. In the other senses, things happen the other way around.

*   *   *

The parts of the brain activated by smells can be visualized by a powerful medical imaging technique called positron emission tomography (PET). PET scanners work by measuring blood flow by means of a radioactive tracer introduced into the bloodstream. Active parts show up as bright patches on the monitor screen. Ivanca Savic and her group at the Karolinska Institute in Stockholm have successfully used the technique to work out how odorants are processed by the brain. To track what happens when you smell a smell, she established four groups each of 12 fit, healthy young men and women aged between 20 and 28, all of whom had a normal sense of smell.[41,42,43] There were no smokers in the groups, and all were right-handed. Women were tested during the second and third weeks of their menstrual cycles (when smell perception is keenest), and none was taking oral contraceptives. The four groups comprised heterosexual men, heterosexual women, homosexual men, and homosexual women. Savic chose to study the sexual orientation of her subjects, as well as their gender, in order to obtain a more comprehensive overview of how brains process olfactory signals. Subjects were presented first with a series of ordinary smells — lavender oil, cedar oil, eugenol (smelling of cloves), and butanol (a sweetly alcoholic smell, used widely in the food industry). Exposure to ordinary

---

[41] Savic, I., and Lindström, P. 2008 PET and MRI show differences in cerebral asymmetry and functional connectivity between homo- and heterosexual subjects. Proceedings of the National Academy of Sciences USA, 105: 9403–9408.

[42] Savic, I., Berglund, H., and Lindström, P. 2005 Brain response to putative pheromones in homosexual men. Proceedings of the National Academy of Sciences USA, 102(20): 7356–7361.

[43] Berglund, H., Lindström, P., and Savic, I. 2006 Brain responses to putative pheromones in lesbian women. Proceedings of the National Academy of Sciences USA 103(21): 8269–8274.

smells resulted in activation of the olfactory tubercle, the piriform cortex, the entorhinal cortex, amygdala, and hypothalamus — in other words the classical olfactory region, irrespective of biological sex or sexual orientation. Men and women process ordinary environmental smells in the same parts of the brain — a hardly unexpected result. It was only when she presented smells known to be detectable at different concentrations by men and women that differences appeared. The special smells used were 'androstadienone', a testosterone derivative found in male underarm secretion, and 'oestratetraenol', an oestrogen derivative found in women's urine.

Heterosexual women exposed to the testosterone derivative showed strong brain activation in the anterior hypothalamus as well as in the piriform cortex. When exposed to the oestrogen derivative, their piriform cortices showed activation, but nothing happened in their hypothalamic regions, meaning that they perceived the female compound much as they would any ordinary smell. Male smell, however, targeted the part of the brain central to sexual behaviour. Similarly, the brains of heterosexual men exposed to the oestrogen derivative, showed activation of the anterior hypothalamus. The activation was not in exactly the same parts of the hypothalamus that were activated when women smelled male extract, indicating that the hypothalamus has some sex-specific regions. When heterosexual men were exposed to the smell of the male derivative, the hypothalamus remained quiet, while the piriform cortex was activated as it is with ordinary smells.

The hypothalamus of homosexual women lit up in response to the oestrogen derivative, but remained quiescent when the male derivative was presented. And when homosexual men perceived the male derivative, their hypothalamic region responded, and similarly remained quiescent to the female derivative. What Savic and her team showed was that while ordinary smells are processed identically by people across a range of sexual orientations, smells that originate in the sex hormones are processed differently by men and women. The hypothalamus appears to be intimately associated with a person's sexual orientation, as well as with their biological sex.

Several parts of the brain show great differences in size between men and women. In particular, the hypothalamus and amygdala of men are twice the size they are in women. The amygdala, in particular, is highly responsive to the male sex hormone testosterone, shrinking in size by a third following castration. The amygdalae of heterosexual women and homosexual men have more connections to other parts of the limbic system in the left side of the brain, while in heterosexual men and homosexual women the greater number of connections is with the right side.

Savic's studies were conducted under artificial laboratory conditions, necessary for operating the complex PET equipment. Are their findings repeated in real life, by real people interacting with one another? Yolanda Martins and colleagues at the Monell Chemical Senses Center investigated the preferences for human body odour according to gender and sexual orientation, and found that axillary secretions from gay men are reportedly less pleasant to straight men, straight women and lesbians than secretions from heterosexual men, heterosexual women and lesbians, but more pleasant to homosexual men.[44] When they compared axillary smells from straight men and straight women, homosexual men reportedly preferred the smell of straight women. Preferences were statistically significant at a high level, giving confidence that the observations reflect reality.

In the light of what we know about how sexual orientation affects smell processing, it's hardly surprising that there are differences in gross brain architecture between homo- and heterosexual men, and homo- and heterosexual women. Heterosexual men and homosexual women have brains in which the right cerebral hemisphere is larger than the left, while homosexual men and heterosexual women have much more symmetrically shaped brains. The neural connections from the limbic system to other parts of the brain in homosexual men

---

[44] Martins, Y., Preti, G.J., Crabtree, C.R., *et al.* 2005 Preference for human body odor is influenced by gender and sexual orientation. Psychological Science 16: 694–701.

resembles far more the pattern found in heterosexual women than heterosexual men, and *vice versa* for homosexual women.

We'll likely never fully understand how the amygdala and hypothalamus interact with one another, and with other parts of the brain to make sense of smells, but we do know they are deeply involved in the process of sexual arousal and the release of sexual behaviour. In mammals, the VN system targets the amygdala directly as its first port-of-call; the main olfactory system targets it only after the sensory impulse has already been read by the piriform cortex, and possibly sent off to the thalamus for rational scrutiny. As we'll see later, this is a difference of absolutely fundamental importance. Lacking a functional VNO, humans are only able to stimulate the amygdala *via* the nose, and not directly as occurs in other mammals.

$$*\quad*\quad*$$

The VNO was noted in the early 1700s by Frederik Ruysch, forensic advisor to the Amsterdam courts, but not described fully until 100 years later. This task fell to the Danish physician Ludwig Jacobson who, for want of other things to do during the British siege of Copenhagen in 1807, examined the anatomy of the mouths and noses of mammals. He described in detail the structure lying above the hard palate in the roof of the mouth embedded in the vomer bone, and in the mid-line of the nose underneath the nasal septum (Figure 5.2). The rather cumbersome name for this structure comes from its anatomical position, though it's sometimes referred to as 'Jacobson's Organ', in honour of its describer. In keeping with modern convention I'll use its positional name.

The VNO consists of a pair of blind-ended pouches, rather like socks, lined with olfactory membrane, from which nerves pass to the brain. These nerves, called the 'vomeronasal nerves', run to the brain in close juxtaposition to, but not joining with, the olfactory nerves. They project first not to the olfactory bulbs, where the olfactory nerves go, but to structures called 'accessory olfactory bulbs' that are tightly attached to the backs of the olfactory bulbs, but are quite independent of them. Leaving the accessory olfactory bulbs, the second VN neurons again travel in tandem with the olfactory neurons, but

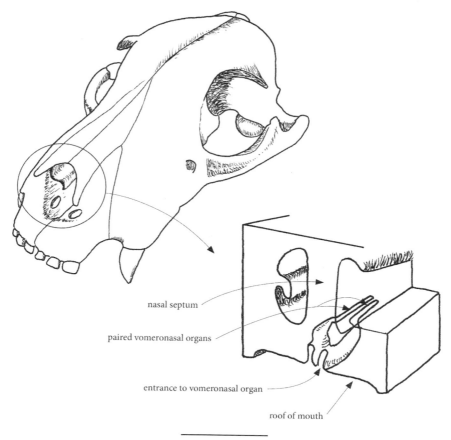

nasal septum

paired vomeronasal organs

entrance to vomeronasal organ

roof of mouth

Figure 5.2

A mammalian skull showing the location of the vomeronasal organ, lying in the hard palate of the roof of the mouth, at the base of the nasal septum.

now things change a bit, because they bypass the piriform cortex and head directly for the amygdala. The VNO nerve projects into parts of the amygdala not targeted by the nose. A side-shoot targets a structure called the 'stria terminalis' — another part of the brain that is very sensitive to testosterone and plays a role in regulating the body's hormone system. The VNO pathway is shown in (Figure 5.3).

The amygdala's main sensory inputs are from the olfactory and visual systems, though it also receives input from the other senses.

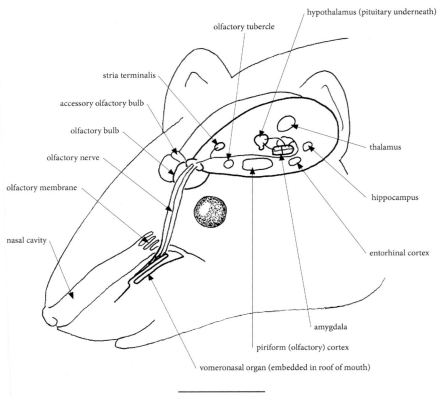

Figure 5.3

Schematic showing the main vomeronasal nerve pathway to the key parts of the smelling brain in mammals. For clarity, the pathway is shown only on one side of the brain.

It's been described as the CEO of the limbic system, because of the power it wields over the hypothalamus. Among its many roles is the coordination of pleasure, as witnessed by the large number of opiate receptors present in it; in fact it contains the greatest concentration of pleasure receptors in the whole brain. In some way that is not fully understood, it's also involved in addictive behaviour, doubtless involving its pleasure receptors. The amygdala contains sex and stress hormones, revealing its connections with sex and stress. Studies using male rats have shown that olfactory and visual sexual signals are critically mediated by the amygdala. Surgical lesions to the amygdala seriously interfere with the ability of male rats to find a mate, or to engage in sexual behaviour, even when receiving strong sexual signals.

In humans, too, the amygdala plays a part in sexual behaviour. Stephan Hamann of Emory University subjected university undergraduates to erotic images of copulating couples, as well as to pictures of men and women in non-sexual situations.[45] At the same time as the pictures were being viewed, the activity of the amygdala was visualized using fMRI technology. Scans showed that the amygdalae of men were more strongly aroused by visual sexual stimuli than were those of women, irrespective of how the subjects later reported they were aroused by the images. Non-sexual pictures elicited equally low amygdala response in both sexes.

As noted above, the crucial difference between the VNO's route to the brain, and the route taken by the nose, is that the VNO bypasses the piriform cortex, with its connections to the thalamus and the rational brain, so its signals aren't able to be interrogated rationally. Because there's no chance for VNO-generated signals to be interrogated by the rational brain, it's not surprising to learn that the VN system is fundamentally committed to the coordination of sex in just about all species of mammals.

*   *   *

The VNO works like this. Chemical substances from the environment enter the organ's blind-ended pouches through one of several mechanisms. Anyone who spends time with horses or goats will be aware of a special face males make during the breeding season, when the lips and tip of the snout are drawn tightly back while the animal makes a short and energetic sucking in of breath. It's sometimes depicted as looking as though the horse is laughing (Figure 5.4).

You can also see it in dogs and cats, and especially well in colonies of mountain goats in zoos and wildlife parks. The strange sucking in behaviour with the drawn-back lips is called '*flehmen*', a German

[45]Hamann, S., Herman, R.A., Nolan, C.L. *et al.* 2004 Men and women differ in amygdala response to visual sexual stimuli. Nature Neuroscience 7(4): 411–416.

(a)　　　　　　　　　　　　　　　　(b)

(c)

Figure 5.4

Flehmen in (a) horse (b) buffalo (c) rhinoceros. The drawn-back lips open the entrance to the vomeronasal organ. (a), (b) and (c) Copyright Ardea Picture Library Ltd.

word that defies English translation. The curled lips ensure that the end of the passage from the VNO to the roof of the mouth is opened up to receive scent. Accompanying this behaviour is a pumping action of blood vessels lying in the organ's floor that slightly reduces its internal pressure, so helping suck in odorants. Later they will reverse the flow, pumping out the last sample in time for a new one to enter. *Flehmen* occurs in the context of sexual behaviour and is typically displayed after the male has nuzzled the perineum of a female, inducing her to urinate. He sucks up some of her urine and exhibits *flehmen* as the odorants in the urine are injected into the VNO. Lower primates (tarsiers, lorises, lemurs) and the New World monkeys (marmosets, tamarins, squirrel monkeys, owl monkeys, and the like) have well-developed VNOs and exhibit *flehmen*, though their behaviour is nowhere nearly as flamboyant as that of horses and antelopes.

*   *   *

Why do animals have a two-part sense of smell — the nose and the VNO? Fish have no VNO, although they have some VNO receptors expressed in the main olfactory membrane, alongside olfactory receptors. This suggests there may be little difference between the two types of receptor. Recall that waterborne molecules do not have to weigh less than 300 Daltons for them to be sensed by aquatic animals, enabling large molecules the size of proteins to be 'smelled'. Air-breathing animals can sample only small, light molecules. This size constraint helps underpin two hypotheses purporting to explain why mammals have a dual smelling system.[46]

The first hypothesis is based on the fact that chemical compounds with high molecular weights lack the volatility to become airborne and thus can't be sampled by the nose. The only way a mammal can perceive a molecule weighing more than 300 Daltons is by 'tasting' it, and this is, in effect, what the VNO does. The second hypothesis holds that the VNO is specially adapted to perceive only pheromones, and is based on the anatomical fact that its nerves project directly to the amygdala and hypothalamus, and not to the brain's primary smell centres. The evidence showing mammalian pheromones, perceived *via* the VNO and used in orchestrating sexual behaviour and reproduction, is substantial, with many well-documented experimental studies reported in the scientific literature. Pheromones have been shown to be involved in sexual behaviour from the advertisement of sexual maturation to the induction of mating, and from acceleration of sexual maturation in others, to the maintenance of pregnancy.

There is a substantial body of experimental evidence to support each hypothesis, but also much that doesn't. For example, in some species of mammals, the VNO is used for detecting non-pheromonal

[46] Boxa, K.N., Dorries, K.M., and Eisthen, H.L. 2006 Is the vomeronasal system really specialized for detecting pheromones? TRENDS in Neuroscience 29. doi:10.1016/j.tins.2005.10.002.

scents; if North American opossums have their VNOs experimentally blocked, they're unable to find food or make normal decisions about what to eat. On the other hand, it's well known that male pigs produce a mating pheromone in their saliva that has a dramatic effect on sows in the ovulatory phase of their cycles, inducing them to adopt a rigid mating posture and inviting the male to mount them. The VNO plays no part here; the behaviour is activated by the main olfactory system, *via* the nose. Other species have a bet each way. In golden hamsters, for example, the VNO is critical for normal mating behaviour in sexually naïve males, but once initial sexual experience has been gained the VNO is no longer used, and the main olfactory system takes over the job of detecting females on heat. These variations on the VNO theme show us what we see time and again in biology, namely that one explanation seldom covers every observation, and that the evolving bodies of animals are like putty in the hands of natural selection. Indeed, experimental work with mice shows that genetic disruption of the main olfactory system results in the VNO adapting to detect environmental smells to which it wouldn't otherwise respond.[47] The evidence is that there is a huge amount of sensory plasticity in how mammals detect the state of oestrus in their partners, with the VNO being essential in the majority of species, but not all.

---

[47] Rodriguez, I. 2003 Nosing into pheromone detectors. Nature Neuroscience 6 (5): 438–440.

Chapter 6

Sex, Smell and 'ADAM'

Sexual reproduction has a simple objective; to ensure that gametes from the male (sperm) meet gametes from the female (eggs) when both are ripe, for only then will fertilization occur, and only then will the genes of both parents be on their way to the next generation. To get there, however, requires careful interaction between the sexes, generally starting with the two sexes coming together, sometimes from far afield. It may involve the male courting the female to diminish her aggression so he can mate, or to assure himself that she is not already pregnant and carrying some other male's offspring. Sometimes courtship hastens the physiological readiness of the pair, a process that might take hours, or days. Without doubt, sex is the most complex thing an animal ever has to do, for if any step in the chain from the initial attraction of the sexes to consummation is out of place, reproduction won't occur, and the opportunity to contribute genes into the next generation will be lost. For long-lived species there may be other opportunities, but for those that live only one year, their single chance will be gone forever.

Like an iceberg's hidden nine-tenths, observed sex is the culmination of a lengthy, multi-stepped chain of physiological events going on inside the animal. Testes and ovaries must be brought to the right stage of development, and the womb readied to receive embryos. The requirement in mammals for the young to be retained within the body until they are ready to be born requires yet more physiological pathways to be switched on, only after fertilization and implantation of the embryos in the uterus wall have occurred. And then come yet more pathways to support lactation and associated maternal behaviour. All of these require precise coordination, so that each step occurs at just the right time. In all vertebrate animals, the coordination control centre for these matters is a structure lying just underneath the brain called the 'pituitary', sometimes known as the 'pituitary body' or 'pituitary gland'. In humans it's about the size of a pea. No trace of a pituitary has been found in any invertebrate animal, though as we'll see, its functions are carried out by other structures.

The word 'pituitary' comes from the Latin 'pituitarius', meaning 'to secrete phlegm', but what it actually secretes is a range of hormones that fire up many of the body's organs. Sex isn't its only responsibility; its hormones control metabolism, growth, stress, blood pressure, water balance and the kidneys, body temperature, pregnancy and birth, and various other things. It may be helpful to imagine it as the central media office of a large organization, represented in our case by the brain, through which the organization's plans are implemented. The pituitary has aptly been described as the body's master gland, because of its influence over just about all the body's physiological activities.

Despite being called by a name that suggests it's a single structure, as the words 'kidney', 'liver', or 'heart' each denote a single organ, the pituitary is a pairing of two quite separate and distinct structures, each with its own role. The part nearest the front of the head, called the 'anterior pituitary', has the job of secreting the hormones that control the body's functions, sending them out in the blood circulation system to be picked up and acted upon by the appropriate organ. The part lying behind it is called the 'posterior pituitary' and is, in essence, a downward extension of the hypothalamus, whose function is to send specific instructions to the anterior part, telling it what messages to distribute. The manner in which the two parts interact to coordinate sex is a product of them both having developed from a patch of olfactory tissue that sits at the front end of the developing embryo.

An early attempt to understand how sex is coordinated in animals was made by the English physiologist David Carlyle, who worked on the insignificant little sea squirt — small gelatinous animals you'll find clinging to rocks in tide pools, or festooning pier piles. He didn't pick these little animals at random for study. As a classically trained zoologist he knew they are neither invertebrate nor vertebrate, but sat at the junction in the evolutionary tree when vertebrates started to differentiate themselves from all other animals. This happened about 520 million years ago, in the Cambrian epoch, so sea squirts ought to be able to give us a picture about how sex was coordinated before the first vertebrates appeared.

We know sea squirts occupy an intermediate evolutionary position because of the presence in their larval tails of stiffening rods that lets them swim, in much the way tadpoles use their tails to swim. We still possess that stiffening rod to this day, though as pregnancy progresses it becomes relegated to the cartilaginous discs that separate the bones in our spines. After the sea squirt larvae have spent some time living among the plankton in the ocean, they use their tadpole-like tails to swim downwards to find a suitable place to live. They become attached to a hard surface and the tail, together with its stiffening rod, degenerates. There's no sign of the tail rod in the adult sea squirt, which spends the rest of its days quietly straining tiny particles of food from seawater. Anatomists in the 19th century recognized the importance of sea squirts in the evolution of vertebrates and as a consequence studied them closely. It turns out they are a lot less boring than you might think (the animals, that is!) because, as so often is the case in evolution, they turn our attention to sex.

Carlyle asked some questions about how sessile sea squirts manage to coordinate their reproduction, for the evolutionary record makes it clear they have been doing it very successfully for half a billion years. Males and females have no sexual organs for the transference of sperm or eggs; instead they release their gametes into the water where random fertilization takes place. The puzzling question is how do they coordinate the timing of their release so the gametes aren't wasted? It turns out they employ processes that reflect what goes on in our pituitaries.

When the larvae are swimming about, the actions of their tails are controlled by a bundle of nerves — hardly a brain as we know it, but certainly a control centre for the larva's activities. After settlement,

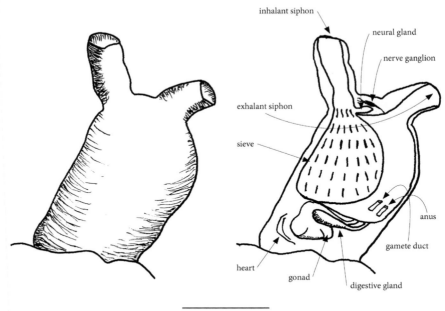

inhalant siphon

neural gland

nerve ganglion

exhalant siphon

sieve

anus

gamete duct

heart

gonad

digestive gland

Figure 6.1

Adult sea squirt. On the right, the cut-away sketch shows the seive, neural gland, nerve ganglion and gonad.

the nerve ganglion, as the bundle is called, reduces in size, but does not completely disappear. As the young sea squirt develops its adult body form, a small pocket develops from the top of the water sieve, which grows out towards the nerve ganglion and comes to lie snugly against it, though remaining separate from it. The pocket is known as the 'neural gland'. Figure 6.1 shows a diagram of an adult sea squirt, cutaway to show its internal structure.

By introducing some fine particles of dye into the sea squirt's inhalant siphon, Carlyle demonstrated that some water was propelled into the neural gland. He deduced from this experiment that the neural gland's role was to sample the water drawn into the squirt.[48] Next, he introduced a hormone extracted from the urine of pregnant

[48] Carlyle, D.B. 1951 On the hormones and neural control of the release of gametes in ascidians. Journal of Experimental Biology 28: 463–472.

women, called 'human chorionic gonadotrophin', or hCG, into the inflow of water, and found that as soon as this was sampled by the neural gland, the sea squirt responded by releasing its gametes into the exhalant water stream, making the surrounding water cloudy. If he injected hCG into the body of the animal, or even directly into the gonads, nothing happened. The response occurred only if the hormone was introduced into the inhalant water flow from where it could access the neural gland and the nerve ganglion. His most telling experiment was to introduce some gametes from another sea squirt of the same species into the incoming water flow, to be met with the release of the test animal's own gametes. When he introduced gametes from a different species of sea squirt, nothing happened — release occurred only in response to gametes of the right species. Carlyle concluded that the neural gland acted as an environmental sensor, like an aquatic nose, passing messages to the nerve ganglion about the sexual state of play of other squirts. When those messages showed that gametes from others of the same species were present in the inhalant water stream, the nerve ganglion responded by sending a hormonal message to the gonads that the time was right for them to release their gametes, and so sexual reproduction was coordinated. (If every sea squirt is waiting for every other squirt to release their gametes first, how is the whole cycle initiated? In all probability there is some leakage from the ripest gonads — enough to trigger the neural glands of others.)

There is some similarity between the neural gland complex in sea squirts and the pituitary in mammals, and one significant difference. The sea squirt neural gland reads signs about the state of readiness in other individuals, and triggers gamete release at the right moment. In the last respect it acts just like the pituitary. The significant difference is that the sea squirt neural gland and nerve ganglion complex acts as both environmental sensor, *and* as the body's master coordinator, whereas the mammalian pituitary doesn't sample the outside environment; it only releases hormonal signals to the gonads. Right across the animal kingdom there are examples of environmental sensors triggering hormonal chain reactions, resulting in mating and fertilization. In every case it turns out that it is the smell organ that senses the environment and passes sex coordination messages to the animal's brain.

There has been much research involving a number of invertebrate species and a fascinating story is emerging. Research on sponges, jellyfish, roundworms, molluscs, earthworms, arthropods, starfish, as well as on sea squirts, shows the occurrence of a master sex hormone called 'gonadotrophin releasing hormone', or GnRH for short. The scientific name for this hormone simply means it's a hormone that feeds the gonads (testes and ovaries) to the point where they can release their own sex hormones — testosterone and oestrogen. The late Aubrey Gorbman of the University of Washington devoted many years of his life considering the evolution of the mammalian pituitary, and concluded that GnRH has been found in every phylum of animals in which it's been looked for, opening up the fundamentally important possibility that GnRH is *the* universal sex hormone.[49, 50] Right across the invertebrate world GnRH is produced by the organs that detect environmental smells. But in humans, and all other air-breathing vertebrates, GnRH is produced by the hypothalamus of the brain and not by the nose. Significantly, specialized GnRH secreting nerve cells from the hypothalamus project right into the posterior pituitary. The posterior pituitary then stimulates the anterior pituitary to act, messages are sent off to the gonads, and the individual is physiologically and behaviourally readied for reproduction (Figure 6.2).

*   *   *

The relationship between what's happening in the outside world and the hormone chain controlling sex in mammals can be understood by considering the embryonic development of both the nose and the pituitary. The anterior pituitary starts its life as a group of cells located at the very front of the embryo, in a patch of tissue called the 'placodal thickening', sandwiched between two nasal placodes that will become the left and right nasal membranes. Not only will this

---

[49] Gorbman, A. 1955 Olfactory origins and evolution of the brain–pituitary endocrine system: facts and speculation. General and Comparative Endocrinology 97: 171–178.

[50] Gorbman, A., and Sower, S. 2003 Evolution of the role of GnRH in animal (Metazoan) biology. General and Comparative Endocrinology 134: 207–213.

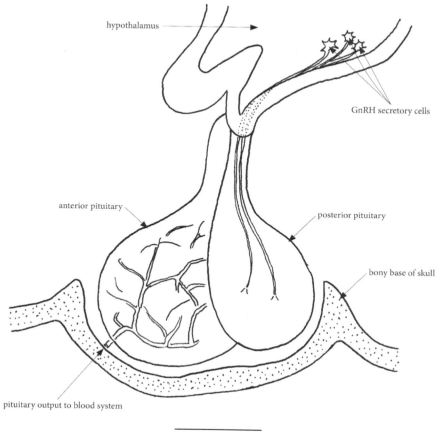

hypothalamus

GnRH secretory cells

anterior pituitary

posterior pituitary

bony base of skull

pituitary output to blood system

Figure 6.2

The anterior pituitary, on the left, and posterior pituitary, on the right. Note the connection between the posterior pituitary and the hypothalamus of the brain.

patch become the anterior pituitary, it will also become the back of the mouth and the upper throat — known as the pharynx. Lying immediately above the nasal and pharyngeal placodes is a third patch, destined to become the hypothalamus. The three patches share this developmental origin and continue their association in adult life, even though they end up in different parts of the head. As the embryo develops, the nasal placodes remain at the front of the head and become the developing nasal passages and olfactory membrane. The

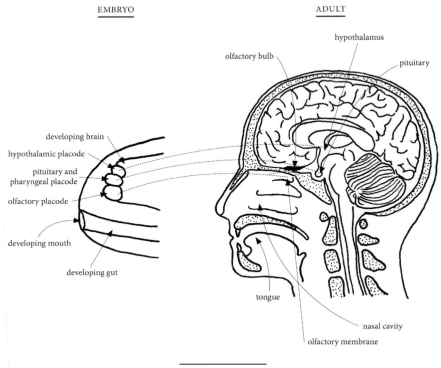

EMBRYO   ADULT

hypothalamus

olfactory bulb

pituitary

developing brain

hypothalamic placode

pituitary and
pharyngeal placode

olfactory placode

developing mouth

developing gut

tongue

nasal cavity

olfactory membrane

Figure 6.3

Embryonic and adult positions of the hypothalamic, pituitary and nasal placodes.
On the left is a detailed sketch of the embroyo's developing placodes, relative to the
developing gut and brain. Their adult positions are shown on the right.

pharyngeal placode migrates rearwards to become the back of the
throat and the anterior pituitary. An open passageway is retained
between them for the first two months of pregnancy, after which time
it degenerates and disappears. The hypothalamic patch migrates rearwards, coming to rest in the floor of the developing brain. Figure 6.3
shows these relationships in diagrammatic form.

The three embryonic neighbours cooperate throughout life to
coordinate sex, even though, in adulthood, they are no longer in
contact with one another. They maintain contact through a chain of
nerve cells, linking the outside world with the gonads in what's known

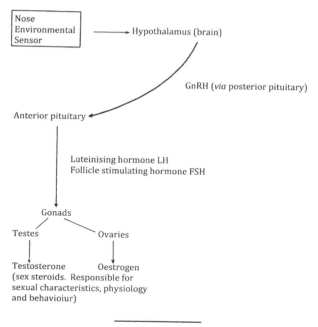

Figure 6.4

The brain-pituitary-gonad link.

as the brain–pituitary–gonad, or BPG, link. Figure 6.4 shows the BPG link in schematic form.

Although air-breathing brings many advantages, a significant disadvantage is that the nose must communicate with the brain through a chain of nerves, rather than directly, as in sea squirts. Ultimately, the chain must release the universal sex hormone, GnRH, otherwise sex can't happen. GnRH is not expressed in the adult nose, but the cells that will later express it in the hypothalamus start their lives in the nasal placode of the embryo. Soon after conception GnRH cells attach to the hypothalamic patch and migrate with it towards the developing brain. If anything should happen to upset the migration of GnRH cells, and they don't arrive in the brain, the BPG link will be broken and sex won't occur. This is what happens to people suffering from what's known as Kallmann syndrome, a condition afflicting about

1 in 10,000 people, with about three times as many men suffering as women. In this condition some mishap occurs during embryonic development to cause the olfactory bulbs of the brain to develop incorrectly, thus blocking the brainwards migration of GnRH cells streaming out from the nasal placode. GnRH cells travel *via* the olfactory nerve and the olfactory bulbs, and the VNO nerves and accessory olfactory bulbs in humans as in all mammals, and if the bulbs fail to develop correctly, the GnRH cells remain tangled up in the olfactory nerve tract just outside the brain, unable to deliver their messages anywhere. If the hypothalamus doesn't receive its GnRH cells early in embryonic development, and thus isn't able to produce the GnRH signals necessary to tell the anterior pituitary that it's time for puberty to commence, sufferers from Kallmann syndrome remain sexually immature with undeveloped testes and ovaries. The failure of puberty is often the first inkling that anything is amiss, though sufferers lack a sense of smell from infancy, or at least have a markedly impaired smelling ability. In our largely visual and acoustic world, it's possible for a poor or even absent sense of smell to be overlooked. Therapy with GnRH can bring about the onset of puberty and the acquisition of secondary sexual characteristics, and even support a near-normal sex life, but it doesn't help the loss of the sense of smell. The migration of GnRH cells from the nose to the brain can only occur through a narrow time window during embryonic development that, once closed, can't be reopened.

In humans, the anterior pituitary responds to the GnRH message produced by the hypothalamus, and sent to it *via* the posterior pituitary, by releasing two hormones into the bloodstream to stimulate the gonads. These are called luteinizing hormone (LH), and follicle-stimulating hormone (FSH) (Figure. 6.4). LH and FSH act on both testes and ovaries, bringing them to maturity, ripening up sperm and eggs for reproduction and, in females, preparing the uterus for pregnancy. As the testes and ovaries mature, these organs start to produce their own steroid hormones, (testosterone and oestrogen) which bring about the physical manifestations of sexual maturity, including beard growth and deep voice in males, and breast development and pelvis widening in females. They also cause hair to grow in the armpits, on the pubis, and around the genitals. Significantly for Adam and Eve, they also switch on the body's apocrine glands.

So, what does all this mean for the role of the human nose in sex? After a lifetime's reflection on the brain, the pituitary and the coordination of sex, Aubrey Gorbman concluded that the principal difference between animals like the sea squirt and humans is that the nose has transferred its hormone secretory function to the hypothalamus *via* its embryonic link to the nose, and it's the hypothalamus that releases GnRH to the anterior pituitary. For its part, the hypothalamus' connection with the smells of the outside world now occurs through nerves running from the olfactory membrane to it. Gorbman further concluded that the evolutionary benefit of this transfer is that it enables the anterior pituitary to respond to a much broader range of incoming nervous stimulations from the brain, and is no longer restricted to respond only to olfactory stimuli. Thus the eyes and ears, as well as touch and taste, can now stimulate the limbic system, so opening up a whole host of ways in which sex can be initiated. The nose is still involved, but now it works in partnership with inputs from the other senses. Cleopatra's seduction of Marc Antony didn't depend only upon the fragrances she, and he, had applied to their bodies, but also upon the soft background music, Cleopatra's carefully applied make-up and Marc Antony's manly looks, their freely traded caresses, and the *haute cuisine* and aromatic wine, all contributing to a multi-sensory seduction.

* * *

More's known about the coordination of sex in mice than in just about any other mammal, because of their widespread use in the development of pharmaceuticals. The mouse is an excellent model species because they breed the whole year round, have a short egg-production cycle of 4–5 days, mature at between 5 and 8 weeks, and can be caged at high densities. It was a decade after the end of WWII, when interest in new pharmaceuticals was developing, that reports started circulating of smells acting as a Svengali-like manipulator of the sex lives of mice. Wesley Whitten, an Australian veterinary scientist working in

Canberra, noticed that female mice, if housed in groups and isolated from any contact with males, suffered upsets to their regular cycles of egg development and release into the fallopian tubes, called 'oestrous' cycles.[51] ('Oestrus' is the correct name given to sexual receptivity, or 'heat'. Its adjectival form is 'oestrous'.) He expected the females to cycle randomly, such that when males were introduced to their cages pregnancies would be established randomly over 4–5 days, as each mouse came on heat. Instead he found that over half the females became pregnant on day 3, with the other half falling pregnant over days 1, 2, 4 and 5. Questioning if the spike in conception might be because of chemical cues emanating from males, Whitten surgically removed the olfactory bulbs of 33 females, and six weeks later examined their uteri, ovaries, and body weights. The experiment showed conclusively that all three weights were significantly lower than those of control females. Additionally, their ovaries showed no signs of the egg-production cycle. When a group of males was subjected to the same surgical intervention, no effect was seen on testis weight, but the accessory gland (an important contributor to semen) and overall body weights were somewhat depressed. When these males were paired with control females that had not been subject to any surgical treatment, almost no mating took place, suggesting that a working sense of smell is crucial for correct sexual functioning in both sexes.

At about the same time as Whitten was working in Canberra, two Dutch scientists, S. van der Lee and L.M. Boot, observed that if female mice were housed together, and in isolation from males, their oestrous cycles started to lengthen, until eventually they ceased cycling altogether.[52] This effect on oestrous cycles is now known as the Lee–Boot effect, named after its discoverers as is the convention in physiology. Wesley Whitten examined this phenomenon and neatly showed that the introduction of a drop of mature male mouse urine into the isolated females' cage, or even some soiled bedding from a

---

[51] Whitten, W.K. 1956 The effect of the removal of the olfactory bulbs on the gonads of mice. Journal of Endocrinology 14: 160–163.

[52] Lee, S. van der, and Boot, L.M. 1956 Spontaneous pseudopregnancy in mice. Acta Physiologica et Pharmacologica Neerlandica 4: 422–444.

male's cage, was enough to restore normal cycling.[53] The effect of male urine odour on oestrous cycles is named the Whitten effect, in honour of Wesley Whitten's contribution to reproductive biology.

At the National Institute for Medical Research in London, Hilda Bruce turned her attention to the Whitten effect and conducted some experiments of her own on oestrous cycles and male odour.[54] She went one step further than Whitten and looked at the effect of the male mouse smell on the establishment of pregnancy in females. She observed that the scent of a male, *other than that of the male with which she fell pregnant*, caused the pregnancy to fail. If exposure to the scent of what she called the 'strange' male occurred prior to the embryos implanting in the uterine wall, pregnancy failure was due to non-implantation. If the embryos had already recently implanted, exposure caused spontaneous abortion. Once her litter was lost, and her oestrous cycle returned to normal, the female could mate with the male that had caused her to lose her litter, and successfully bear a litter to term. The effect of 'strange male' scent on early pregnancy in mice is now known as the Bruce effect.

In the mid-1970s, after these important discoveries had been made and interest in vertebrate pheromones was gathering pace, John Vandenbergh and his colleagues at North Carolina State University looked at puberty, and what brings about the onset of a female mouse's first oestrus.[55] He observed that puberty occurred earlier if young female mice were exposed to the smell of urine from mature males as they grew up, and delayed if they were exposed only to the urine smell of mature females. He showed that the flow of GnRH from the hypothalamus to the pituitary was stimulated by the smell of males and inhibited by the smell of females — an effect that now bears Vandenbergh's name.

---

[53] Whitten, W.K. 1956 Modification of the oestrous cycle of the mouse by external stimuli associated with the male. Journal of Endocrinology 13: 399–404.

[54] Bruce, H.M. 1959 An exteroceptive block to pregnancy in the mouse. Nature 184: 264.

[55] Vandenbergh, J.G., Whitsett, J.M., and Lombardi, J.R. 1975 Partial isolation of a pheromone accelerating puberty in female mice. Journal of Reproduction and Fertility 43: 515–523.

Rather less attention has been paid to the link between the nose and sex in male mice than in females, perhaps because Whitten's early work had not shown as dramatic an effect as occurs in females, and perhaps because the pharmaceutical industry was showing interest in developing a contraceptive pill focusing on the oestrous cycle, rather than on any part of male reproductive physiology. Whitten had shown that the testes weight of male mice were not affected by loss of their olfactory bulbs, but that there was a depression in the weight of the accessory glands. It took until the late 1970s before the effect of female odour on males was demonstrated. Frank Bronson at the University of Texas showed that male mice experienced a sharp rise in the level of luteinizing hormone produced by the anterior pituitary when exposed to the urine of a female mouse in the ovulatory phase of her cycle, but not in the non-ovulatory phase.[56] The rise in luteinizing hormone stimulates the males' testes to produce a surge of the male sex hormone, testosterone, leading to restitution of the accessory glands, increased scent-marking behaviour, energetic courtship behaviour, and ultimately mating with the female while she is fertile (Figure 6.5). And in one of those delicious parallels, so often encountered in biology when you take a broad approach, the hormone David Carlyle used to elicit gamete release in sea squirts (human chorionic gonadotrophin hCG) is the same as luteinizing hormone.

The relationship between the nose and the VNO system in the coordination of sex in mice is much clearer than it is in some other mammals. Sex in mice is tightly controlled by signals received by the brain from the VNO. Many studies have confirmed that removal of the VNO, or the accessory olfactory bulbs or VNO nerves, results in the failure of reproduction. Even though Whitten was not primarily interested in the VNO when he conducted his seminal studies on mouse sex, he had almost certainly removed the accessory olfactory bulbs from his experimental mice by accident, along with the olfactory bulbs. Other mammals show a similar response to loss of the VNO. In female sheep, for example, the scent of a mature ram brings

---

[56]Bronson, F.H. 1979 The reproductive ecology of the house mouse. Quarterly Review of Biology 54: 265–299.

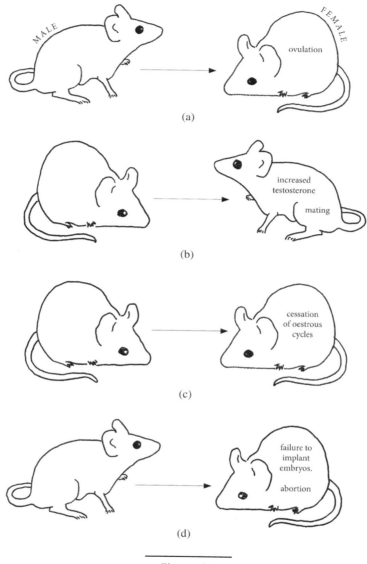

Figure 6.5

The effects of smell on reproduction in mice.
(a) effect of adult male on adult female
(b) effect of adult female on adult male
(c) effect of group housing of females
(d) effect of the smell of a 'strange' male on a recently impregnant female.

about a surge in luteinizing hormone bringing ewes closer to sexual receptivity. The effect is not seen in ewes that have had their VNOs surgically removed.

In some mammals, however, it's the nose and not the VNO system that initiates sex. In pigs, the scent of a boar's frothed up saliva induces sexually receptive sows to adopt a specific mating posture called 'lordosis', in which the back is arched downwards to signify they are fully receptive and ready to mate. The behaviour is retained in sows that have had their VNOs removed, indicating the VNO is not involved in how they perceive the smell. In the golden hamster, to take another example that runs counter to the trend, sexually receptive females advertise their condition by depositing a vaginal secretion on the ground. Males perceive these signals with their VNOs and initiate mating behaviour. Once a male has had a first sexual experience, facilitated by a functioning VNO, his VNO can be surgically removed and he is able to use his main olfactory system for all future detections of receptive females. If a sexually naïve male has his VNO surgically removed, he doesn't respond to sexually receptive females. It's almost as if the VNO acts as a set of training wheels, able to be discarded only in the face — or, rather, smell — of experience.

* * *

While Wesley Whitten was the first scientist to demonstrate experimentally the existence of a link between scent and sex in a laboratory mammal, an association between the nose and sex in humans had been noted since the time of Hippocrates. Contemporary writings inveighed that men shouldn't have sex when they have a cold, as nasal congestion was sure to follow. Heliogabalus, Emperor of Rome around 200 BC, known for his licentiousness and sexual depravity, had biographers noting he'd admit 'to the companionship of his lustful practices' only 'those who were *nasuti,* i.e., who possessed a comeliness of [the nose]'. Further evidence for an understanding of a naso-sexual relationship was that nasal amputation was a widely-practised penalty for adultery throughout the Romano-Grecian and Arab worlds.

John Mackenzie, a nasal surgeon from Baltimore writing in 1884, was the first physician to take seriously the idea that a relationship might exist between the nose and the genital organs in humans, basing his observations on the anatomical similarity between erectile tissues of the nose and the penis.[57] He specialized in treating women suffering nasal congestion during menstruation, and during a bride's honeymoon — the so-called 'bride's cold' — and girls who endured heavy nosebleeds during puberty. Through this work he put intellectual flesh on an emerging theory of a naso-genital pathway. His work heralded a surge of interest in this field that flourished particularly when Wilhelm Fliess, a Berlin doctor and colleague of Sigmund Freud, developed his theory of 'naso-reflex neurosis'.[58] Fliess' theory held that a huge range of conditions, including back pain, digestive upsets, and almost every menstrual dysfunction, could be treated by intervention at various so-called 'genital zones' in the nose. These were specific locations on the erectile membrane lining the nasal cavity. Interventions included surgery in extreme cases and the topical application of cocaine, or cauterization of the nasal membrane, for less severe presentations. The dawn of the 20th century saw an end to these practices, and with it the study of the association between the nose and the reproductive system languished. Only recently, with increased understanding of the body's endocrine system and how it integrates every aspect of our lives, has the theory resurfaced. Within the last few years, the Journal of the Royal Society of Medicine has published a few articles on sex and the nose, with at least one specialist commenting that the profound effects of chronic rhinitis (nasal congestion) on a sufferer's quality of life may be significantly improved by nasal surgery, and that he would not be surprised if diminished libido and sexual performance were noted as consequences of the condition.[59]

---

[57] Mackenzie, J.N. 1884 Irritation of the sexual apparatus as an etiological factor in the production of nasal disease. American Journal of Medical Science 87: 360–365.
[58] Fleiss, W. 1893 Die Nasale Reflexneurose. Bergman, Wiesbaden.
[59] Chester, A.C. 2007 The nose and sex: the nasogenital reflex revisited. Journal of the Royal Society of Medicine 100: 489–490.

The naso-reflex neurosis theory is based on the observation that the lining of the nasal cavity is equipped with patches of erectile tissue, in addition to the sensory tissue. When Fliess examined the noses of his patients he saw that usually either the left or right side of the nose was engorged with blood, causing the airflow through that nostril to be restricted. As we've seen, what Fliess was observing was the normal expansion and contraction of the nasal swell bodies and associated respiratory tissue. But in the context of 19th century Vienna, awash with Freud's focus on sex and sexual repression as the root of all ills, the superficial similarity to sexual tumescence was too much for Dr Fliess to ignore.

What Fliess didn't know was what causes erectile tissues to engorge. Arteries and arterioles are lined with smooth muscle, the tone of which is maintained by certain enzymes. If production of the enzymes is inhibited, the muscle layer relaxes and the arteries balloon out under arterial blood pressure. While the enzymes responsible for maintaining muscle tone are quite restricted to the part of the body in which they work, relaxation of artery tone in one part of the body is not without some effect elsewhere. A commonly reported side effect of the erectile dysfunction drug 'Viagra' is nasal congestion, indicating that the drug's effect isn't completely limited to the *corpus cavernosum* of the penis. Back in the nose, the nasal cycle is repeated eight times a day, driven by a periodic production of smooth muscle relaxant enzymes that has nothing to do with sex.

*   *   *

Humans, being human, have long wondered if their sex lives might be influenced by pheromones, as in mice and most other animals. The concept of insect sex pheromones had been established six decades ago as chemical messages produced by one individual that brought about a *specific effect* on a receiver, in much the same way as pig or dog insulin brings about a specific effect in humans to correct a dysfunctional pancreas. The idea that all animals might have pheromones gained enormous traction in the 1970s and 1980s, and the term came to be applied far more widely than the original definers intended.

Richard Doty, a Philadelphia-based scientist with over four decades of experience in olfactory research behind him, has concluded that there's no evidence that mammalian, and particularly human, pheromones exist.[60] The word 'pheromone' is bandied about quite indiscriminately in the scientific, and especially in the general literature, without proper reference to its original definition. In all studies of putative mammalian pheromones, and particularly in those involving humans, there's no single unequivocal example of a scent or odour produced by one individual eliciting a stereotypic behavioural response that can be compared with the innate responses to pheromones seen in insects. Doty's forensic analysis of the scientific literature shows that the mass media has fuelled a 'junk-science industry' of perfumes, soaps, and cosmetics, inducing us to believe that by using this or that product, specific behavioural outcomes will be achieved. The peddling of so-called human pheromones stands alongside the 19th century's fads of nasology and phrenology in terms of highlighting that the public wants to believe more in mystery than in what can be rationally explained. The fact that there is no proof of the existence of pheromones acting according to the original definition of the term doesn't mean that scents and smells play no part in human sexual behaviour. It means only that cause and effect cannot be demonstrated for a particular behavioural response to a particular smell.

The astute reader will keep in mind that the term 'pheromone' is regularly used inappropriately, when some other term to identify a biologically active smell would be more appropriate. As no other term to embrace the wide range of biologically active scents employed in social interactions has been proposed, it's likely that the term 'pheromone' will continue to be used, despite its considerable shortcomings.

* * *

With Doty's warning in mind, you'll be inundated with almost half a million hits if you search on the internet for the term 'pheromone

---

[60] Doty, R.L. 2010 *The Great Pheromone Myth*. Johns Hopkins University Press, Baltimore.

perfume'. The idea that humans have pheromones seems deeply entrenched and attractive. So what can be said about the explosion in sales of so-called pheromone perfumes over the last decades, if there isn't a shred of evidence that they work? The claims made by manufacturers for the efficacy of bottles of pheromone perfume costing $100 and more, are hard to substantiate through careful evaluation of the scientific literature. Take the following claims from the manufacturers' advertisements, for example:

> 'Pheromones are like a light switch to attract women'
> [when you wear them] 'you subconsciously stand out like a soar (sic) thumb to women.'
>
> 'Pheromones are naturally occurring chemicals that send out subconscious scent signals to the opposite sex that trigger very powerful sexual responses.'
>
> 'Our pheromone perfumes can give you a boost of confidence in the boardroom and the bedroom' (from an advertisement for a pheromone perfume to be worn by women).

Claims such as these suggest that a dab from a bottle is pretty much guaranteed to release sexual behaviour in the perceiver, just as happens when female silkworm moths emit a mating pheromone, or when female mice encounter the soiled bedding of mature males, or when an ovulating sow sniffs the saliva of a rutting boar.[61]

Artificial scents can, and do, boost a wearer's confidence, which certainly helps in developing and maintaining social relationships that may lead to particular desired outcomes, but scientific evidence for a releaser effect, *sensu stricto*, is convincingly lacking. Most species of

---

[61] Boar saliva odour was one of the first sex releaser pheromones discovered and it led to the commercial development of 'Boar Mate' — a spray containing androstenone, the active ingredient in the saliva of mature boars. Pig farmers are able to separate sows ready for artificial insemination by spraying 'Boar Mate' on their noses, identifying those ready by their immediate adoption of a rigid straight-backed mating stance. Internet blogs reveal that 'Boar Mate' has been tried by many men, all of whom report its pulling power leaves something to be desired. Perhaps it has something to do with the smell of pigs!

mammals that produce releaser pheromones have a VNO, something we don't. Our noses smell the scent, but we remain in rational control of our responses.

The fact that humans don't have functioning VNOs doesn't mean we can't, and don't, respond to human scents; it means only that we lack the instinctive behavioural responses that the VNO's hard-wired links with the amygdala and hypothalamus facilitate, because everything smelled by the nose is screened by the rational brain. All evidence points to sexual smells in humans acting subtly through the main olfactory system, and playing a much lesser role than visual and other stimuli in the elicitation of sexual behaviour.

If you are interested in reading more about the science underlying claims made by purveyors of pheromone perfumes, a White Paper prepared for the Fragrance Foundation in 2009 by Charles Wysocki, George Preti and others at the Monell Chemical Senses Center entitled *Human Pheromones: What's Purported, What's Supported* is well worth reading.[62]

*   *   *

Although the evidence for adult humans not possessing active VNOs is strong, there are many who believe otherwise. The debate about a human VNO has raged for decades, supported mainly by those with an interest in the perfume and cosmetics industries. The fact is that embryonic humans show the development of a VNO, complete with what appears to be a sensory lining, but in the vast majority of people it has regressed to nothing by the time of birth. Its peak of development occurs in the 20th week post-conception. In very few people, however, its structure is discernible much later in life and may even give the appearance of being functional.

There are three distinct pieces of evidence indicating that humans lack a functioning VNO. The first is that, despite intensive searching, no trace of accessory olfactory bulbs, nor VN nerves projecting from

---

[62] Available at: www.senseofsmell.org/research/C.Wysocki-White-Paper-Human_Pheromones.pdf.

them to the amygdala, have ever been found. Given the centuries of close anatomical study of human bodies it's inconceivable they would have been overlooked. Even if a VNO persisted into adulthood, and even if it appeared to be fully functional with a well-developed sensory membrane, the absence of nerves dictates it couldn't function. The second piece of evidence comes from genetics. Researchers into olfactory receptor genes have identified two families of G-protein receptor cells expressed in the VNO, similar to those in the main olfactory system. They're called V1R and V2R genes. Janet Young and her associates in Seattle have shown that there's extremely high variability in the genes of the VNO, and about the only thing that can conclusively be said is that if a species has a well-developed VNO, it will likely have a large number of intact VNO genes, and *vice versa*.[63] All V2R genes and all but five of the V1R genes are inactive in humans, unable to express VNO receptors. The molecular evolutionary genetic data about the five apparently active genes strongly suggests they are non-functional; two of them are undergoing the sorts of changes that lead to inactivation, and there's no indication of active selection working on the other three, as would be the case if they were functional. The conclusion must be drawn that there is no evidence of the existence of active V1R genes in humans.

The third piece of evidence comes from an examination of the transduction pathway between the VNOs' receptors and nerve impulse generation. This requires different genes to those performing a similar function in the main olfactory system. The VNO requires a functioning gene named 'TRP2' to open an ion channel to let odorants trigger the VNO's receptors. If that gene isn't functional the channel won't open, and the VNO will not work. Experiments with laboratory mice in which the TRP2 gene has been artificially disrupted result in dramatic disruption to social and sexual behaviour, and have shown it to be essential for correct VNO function. Male mice lacking TRP2 show a lack of male–male aggression and don't seem interested in defending a territory — a defining feature of male mouse behaviour.

---

[63] Young, J.M., Massa, H., Hsu, L, *et al*. 2010 Extreme variability among mammalian V1R gene families. Genome Research doi : 10/1101/gr.098913.109.

They appear unable to see off an intruder, or to form a dominance hierarchy when several are housed together. Female mice are normally far less aggressive than males, showing aggression only when they are lactating. Female mice lacking the gene, however, don't display even this small amount of aggression. Males lacking the TRP2 gene are able to mate normally with females, but they also attempt to mount other males instead of fighting with them, suggesting the TRP2 gene may be involved in discerning gender identity as well as detecting when a female comes on heat. Despite the weight of evidence against the existence of a human VNO, popular magazines and some perfume manufacturers persist with the falsehood that the human sense of smell functions as it does in other mammals. Fortunately for us, it doesn't. What comes next is the key to understanding how Adam's and Eve's sense of smell has contributed, in a fundamental way, to the making of humankind.

*   *   *

Two scientists at the University of Michigan, Jianzhi Zhang and David Webb, examined the TRP2 gene in humans, gorilla, chimpanzees, and monkeys (African baboons and guereza monkeys), as part of a study of the evolution of the VNO.[64] Their work took on a far greater significance than they had anticipated when they demonstrated that a mutation was present in the TRP2 gene of humans, apes, and monkeys from Africa (called 'Old World' monkeys), but not in monkeys from the New World (South American tamarins, saki, squirrel, owl, and spider monkeys). The mutation, which they nicknamed 'ADAM', effectively neutered the VNO's transduction mechanism, yielding the VNO unable to send any message to the brain. Zhang and Webb showed 'ADAM' arose after hominids (humans and the great apes) and Old World monkeys separated from the rest of the monkeys, which is known from the fossil records to have occurred around 35 million years ago. Closer examination of TRP2 in hominids and Old World

---

[64] Zhang, J., and Webb, D.M. 2003 Evolutionary deterioration of the pheromone transduction pathway in catarrhine primates. Proceedings of the National Academy of Sciences USA 100(14): 8337–8341.

monkeys showed there to be only a single protein change between them and the rest of the monkeys, suggesting that 'ADAM' appeared shortly before humans and the great apes split from the rest of the monkeys. The fossil record tells us this event happened about 23 million years ago, providing a timescale for the mutation having arisen sometime between 35 and 23 million years ago, prior to the start of the Miocene epoch and our ancestors' adoption of a communal lifestyle. The acquisition of 'ADAM' was a defining moment in human olfactory evolution.

Further evidence that 'ADAM' successfully did the trick of switching off the VNO came from studies in which Zhang and Webb looked at how many other changes there were in the TRP2 gene, and they found there to be five. Genes that have a function are constantly under selective pressure and show many mutations, but if a gene has no function, and is effectively freewheeling and not being acted upon by natural selection, mutations occur much less frequently. Five mutations in 23 million years accords with what would be predicted for a gene placed under what is called 'neutral evolutionary pressure' for that length of time. All the evidence suggests that the human VNO system was 'decommissioned' by evolution when the hominids split from the remainder of the Old World monkeys, with the result that the physical VNO became vestigial, like so much other anatomical baggage we carry around as a legacy of our evolutionary past.

*　　*　　*

Around the time that 'ADAM' arose, another transformational evolutionary development was taking place in the primates that bore directly on the detection of oestrus. It was the acquisition of full colour vision. It's important to stress there's no genetic evidence to link the loss of VNO functionality with the acquisition of colour vision; the two phenomena haven't been shown to be cause and effect. It's also important to realize that it's impossible to ascribe absolutely accurate timings for genetic changes, so timescales must be interpreted with caution. It was only the Old World primates that acquired the ability to see the world as we do, in what is called 'trichromatic', or three-colour vision; to this day New World primates see life differently.

Old World primates have three genes in the retina of the eye that encode for proteins known as 'opsins'; one that contains pigments sensitive to frequencies of light in the blue range, one in the green range, and one in the red. Genes for the green and red opsins are located on the female sex chromosome — the X chromosome — and the gene for blue is carried on a non-sex chromosome. This would mean that males, who only have a single X chromosome, would carry either a green or a blue gene, and would be green/blue colour-blind. But a process known as 'gene duplication' occurred in Old World primates shortly after they split off from the rest of the primates, in which genes for red and blue occur together on each X chromosome, ensuring that males, as well as females, could perceive the three primary colours. New World monkeys on the other hand, lacked the gene duplication. Only female New World monkeys have the same sort of trichromatic vision as we do, because only they have two 'X' chromosomes, each chromosome carrying a single opsin gene. Males are dichromatic because they have only a single 'X' chromosome and so carry only a green *or* a red opsin, but not both. They're congenitally red-green colour-blind. The full-colour vision enjoyed by our Old World ancestors conferred considerable advantages and was pressed into service by natural selection in two of life's most important activities.

*   *   *

Colour vision evolved in primates at a time when their diet consisted almost exclusively of leaves, fruit, nuts, and the like. Twenty-three million years ago, the primate ancestors of humans were living in forests in which the detection of young leaves and ripe fruit amid dark, mature foliage would have been made much more efficient with the ability to distinguish the reds, yellows, and greens that are the predominant colours of such foods. Colour vision would have brought considerable nutritional benefits.

The acquisition of colour vision provided a visual pathway for females to signal their state of oestrus, to compensate for the loss of the neutered VNO smell pathway. Ovulation advertisement in Old World primates frequently takes the form of exuberant, colourful genital displays, accompanied by sometimes quite massive swelling of the genital region (Figure 6.6). New World monkeys don't show colourful genital displays, because without the ability to see things in colour there would be no advantage. African baboons have brick-red swellings, highlighted with white patches; mangabeys from tropical African forests exhibit fleshy pale red swellings, while female chimpanzees have white swellings. Gorillas and orang-utans have no visible swelling, though in gorillas the external labia sometimes become red and swollen. In gibbons, labial swelling is common. The sexual swellings of baboons can account for 14% of a female's body weight, and in some macaque monkeys even more.

As we've seen, the need for visual sexual signalling in the primate offshoot that gave rise to humans was extinguished when our ancestors became collaborative hunters. Adoption of an upright gait shielded the genital region from view, making it unlikely there would have been any selective pressure for the development of colourful genital swelling and displays. But for our Old World relatives, left behind in the forests, visual displays took over after the VNO signals were interrupted, and have flourished.

Theories abound about how sexual swellings and colourations have evolved, but while interesting, aren't central to the story of the human nose. The interested reader is referred to the work of Charles Nunn for detailed hypotheses.[65] What interests us is that they exist at all. Without the ability to see things in colour, there would be no advantage in sexual displays being colourful. Dramatic and colourful displays send immediate signals that can be perceived from far away, and it was this advantage that fixed the phenomenon of colour vision in the gene pool of Old World primates. What the advent of colour vision did for Old World primates was to provide an information channel for purposes of mate choice *additional to the scent channel*. Only when visual communication had been established could 'ADAM' interrupt the VNO transduction process, decommissioning the VN system without shutting down the coordination of sex. The loss of

---

[65] Nunn, C.L. 1999 The evolution of exaggerated sexual swellings of primates and the graded-signal hypothesis. Animal Behavior 58: 229–246.

(a)

(b)

(c)

(d)

Figure 6.6

Examples of female primate sexual swellings. (a) and (b) sacred baboons; (c) chimpanzee; (d) mangabey monkey (a), (b), and (c) Copyright Ardea Picture Library Ltd; (d) Copyright Dr Tom Struhsaker.

the VN system meant far less for polygynous gorillas, orang-utans, and multi-male dominated chimpanzees, than it did to the only primate that lived gregariously and monogamously.

* * *

So, we have the remarkable occurrence of two monumental gear changes in the evolution of Old World primates, not necessarily occurring simultaneously but likely occurring within a few million years of one another. A few million years may seem like a long time when we consider *H. sapiens* has been distinct for less than half a million years, but it isn't long in the context of primate evolution. The acquisition of colour vision brought advantages to the detection of oestrus and the appearance of 'ADAM' ended the functionality of the VNO. Together they combined to change the sensory world of Old World primates in the most fundamental way. Because the BPG link can be stimulated by senses other than just the sense of smell, better vision enabled more complex visual stimulation to occur. The absence of a functional VNO didn't shut down the role of smells in sexual attraction and coordination; it merely changed the neural pathway by which sex smells reach the brain.

The significance of 'ADAM' to the way our societies are built cannot be underestimated. We remain the smelliest of all the primates, but 'ADAM' has ensured that responses to the smells of others are always subject to rational analysis and a degree of control. Artificially made perfumes and scents only remind us of our mammalian origins; significantly, they do *not* remind us we are humans, for to do so could potentially threaten social order. In this context, it's easy to see how a strong odour culture developed as human societies became more organized and structured. Artificial perfumes derived from sources other than from the human body and including sexual lures of animals and plants, are safe from the possibility that their use might release specific human sexual behaviours.

Chapter **7**

The Scent of Humankind

When Arthur was King of England, the Cornish giant Thunderdell stomped the land claiming he could tell an Englishman by his smell and would rid the country of the foreign scourge.[66] Until the late Middle Ages, all townsfolk in Europe smelled — the English probably no more, nor less, than anyone else. Access to basic hygiene was restricted to the nobility and even then the tub was used sparingly. During their occupation of Britain, Romans bathed as often as they liked in heated bath houses, paying the same meticulous attention to personal hygiene as they did when Cleopatra met Marc Antony, but for ordinary people living in cramped quarters with neither running water nor soap, bodily hygiene ranked pretty low on their daily agendas. The streets were awash with untreated sewage, and garbage and putrefying filth were everywhere. Little wonder the gentry carried posies and nosegays when their business took them out of their grand dwellings — though history reports their quarters didn't smell too good, either! It took until the mid-19th century before sanitary advancement brought cleanliness to the people, and killed city stenches, but until then people just stank.

Setting aside the stink of faeces and putrefaction, what do humans smell like and, for the sake of Thunderdell's claim, does one group of people smell differently to another? Do we like how we smell? Does

---

[66] The giant Thunderdell in *The History of Jack and the Giants*. Newcastle (UK) 1711.

Adam like Eve's, and *vice versa*? Can we read anything into the smell of others, as do mice, moths, and the remainder of creation?

What people smell like is influenced by lots of factors including diet, wellness, age, attention to dental and bodily hygiene, state of mind, and — perhaps surprisingly — even how beautiful they are. Diet is a major determinant of body smell. Big meat eaters have stronger body odours than vegetarians, and people whose diet is primarily fish have a discernibly fishy smell. Strong spices have an effect on body odour; the pungent essence of garlic is excreted out of the body through every pore in the same form as when it was eaten, and smells much the same on its way out as on its way in. In medieval times, garlic's strength was used to warn off the bad spirits that purportedly caused illness, though it did little to abate the great plagues that scourged Europe throughout the Middle Ages.

An individual's state of health is another important determinant of body odour. Many diseases are associated with their own specific smells, as noticed by the Arabian physician Avicenna, and later by Galen, Hippocrates, and the Greek physicians. Typhoid smells of freshly baked bread, scrofula of stale beer, diabetes of nail varnish remover (or pear drops), and rubella of chicken feathers. Experienced doctors make use of these telltale signs in diagnoses. Some very rare genetic abnormalities also cause the body to smell strongly. One such condition is trimethylaminuria, or TMUA, in which the body is unable to break down a chemical compound called trimethylamine which occurs in fish and strong meats, and from an overproduction of gut bacteria that make the compound. The result is that a smell of rotting fish leaks out of the body in urine, sweat, and breath, causing intense embarrassment and much psychological suffering. Only a few hundred cases of TMUA are known worldwide, and while the condition's genetic basis is understood, the only treatments available are to carefully manage diet to diminish the intake of foods that contain trimethylamine, or are broken down into it.

While diet, state of health, and bodily hygiene all have a marked effect upon body odour, batteries of apocrine scent glands are mostly responsible for how a person smells. We are born with one apocrine scent gland opening into the follicle of each hair. They remain quiescent until they are switched on at the onset of sexual maturity. Their lifetime's activity starts just as the body signals the arrival of puberty. Once switched on, they continue working right into old age, long after the production of sex hormones has waned.

Something very interesting happened to sweat glands during mammalian evolution because in many species it's the apocrine glands that produce functional sweat to cool the body. This is so in horses and cloven-hoofed animals, but not in humans. In the line of evolution that gave rise to humans, true sweat glands took on the function of cooling the skin, leaving apocrine glands free to take up their new function of producing smell.

The densest aggregations of apocrine scent glands are clustered in the parts of the body that retain dense hair growth, *viz* the armpits (axillae), the pubic region, around the anus, in the umbilical region, and around the mouth (Figure 7.1). They are found elsewhere on the body but in much lesser densities, including around the areolae and the labia minora in women, and the foreskin in men. Women have more apocrine glands than men and Asians have fewer than Africans, Indians, or Caucasians. In the axilla, up to 100 apocrine glands pack each square centimetre. From an evolutionary point of view, armpits are central to our understanding of much about Adam and Eve's noses, and about what makes us human. We'll look at their evolution in the next chapter.

*　　*　　*

An enormous amount of research has been conducted on human body odour, supported and encouraged by commercial companies making soaps, deodorants, and other personal hygiene products. The size of the deodorant market is huge. In 2011 Unilever, and Proctor and Gamble together turned sales of deodorants of almost $2 billion, with Colgate-Palmolive, Dial Corp, and Revlon hotly in pursuit. World business reviews predict annual growths in the deodorant market of around 10%, with growth up to four times higher in India

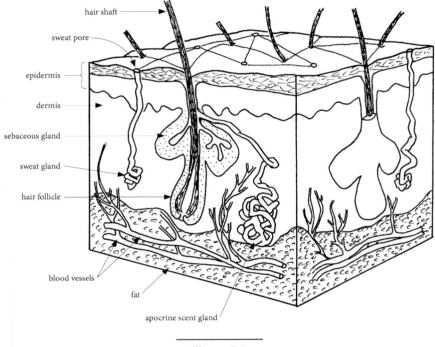

hair shaft

sweat pore

epidermis

dermis

sebaceous gland

sweat gland

hair follicle

blood vessels

fat

apocrine scent gland

Figure 7.1

A close-up view of human skin, showing hairs emerging from hair follicles. Sebaceous glands, which make the hair oily, open into each follicle. True sweat glands open onto the skins surface. Into the top of each hair follicle opens the duct of a long, coiled apocrine, or scent gland.

and the Middle East. Commercial interest in what makes humans smell of humans has never been higher.

Boiled down to chemical basics, healthy humans smell of a mixture of volatile organic fatty acids, together with small quantities of odorous steroids. Molecules of the acids are quite small, having between 6 and 11 carbon atoms, and come in various configurations. They're volatile and readily carried on the air. Work over many years conducted by George Preti and his colleagues at the Monell Chemical Senses Center in Philadelphia has revealed the composition of human body odour.[67]

---

[67] Pretti, G., and Leyden, J.J. 2010 Genetic influences on human body odor: from genes to the axillae. Journal of Investigative Dermatology 130: 344–346.

Of the 15 or so chemical compounds that make up the typical human smell, the heaviest molecule has a weight of 186 Daltons — well below the 300 Dalton limit for volatility. The most abundant component has the chemical name 3-methyl-2-hexanoic acid, a compound that human sniffers describe as having a urinous, musky, sweaty, burnt, or woody smell.

Although this compound is one of the main sources of characteristic human odour, the outpouring of the apocrine glands is initially without smell. Left in the warmth and humidity of the axilla, the legions of skin bacteria living there break down the secretion to produce the familiar cocktail of scented acids, and after about six hours the fully developed characteristic smell emerges. The bacteria responsible for breaking down apocrine gland secretions belong to two main types — rod-shaped coryneform bacteria and spherical micrococci; the proportions of the two types in men and women are different, but both types occur in both sexes. Axillary hair provides a huge surface area on which the bacteria can perform their work, as well as to act as a wick to disperse the scent when the arms are raised, and the axillae are exposed to the air. In qualitative terms there are few differences between the odorants produced in the axillae of men and women, though men tend to produce odours reported to be of a sharper intensity.

Along with the essential acidic ingredients of human odour, the axillary secretions of men and women contain small quantities of scented steroids derived from the sex hormone testosterone, and named 'androstadienone', 'androstenone', and 'androstenol'. They're much more abundant in men than women. With musky, slightly urinous smells not unlike that of sandalwood, Preti and his associates indicate that recognizable underarm smell comes not from the small quantities of steroid chemicals, but from the abundant acids. Androstadienone, androstenone, and androstenol are far from being biologically inert, with humans showing marked age-dependent sensitivity to androstenone and androstadienone. In pigs, androstenone is the mating pheromone eliciting sows to adopt the mating posture.

Before reaching puberty, children of both sexes are very sensitive to the smell of androstenone and androstadienone, but as they become sexually mature young men become less sensitive, with almost 20% not being able to smell it at all. Young women, on the other hand, retain sensitivity to both steroids, though they experience a cycle of increasing and decreasing sensitivity as the menstrual cycle goes through its follicular and ovulatory phases. It's been postulated that decreasing sensitivity in young men as they mature may be because their blood system remains flooded with testicular steroid derivatives at a high concentration.[68]

*   *   *

There's a strong genetic influence on the characteristics of human body odour, just as there's a strong influence on skin colour, facial features, stature, and everything else that marks us as our parents' children. There is a clear genetically determined difference between the mild body odour of Asians, and the fierce smell of Africans and Caucasians. Annette Martin and colleagues have identified a gene that's responsible for the production of a protein molecule which carries the important axillary odorant 3-methyl-2-hexanoic acid from its production site deep in the skin to the surface, where bacteria can get at it.[69] Martin and her team showed that there's a small mutation to the gene which, when occurring on both chromosomes, blocks the production of a transportation protein. If a person has two blocking alleles, as such mutations are called, very little apocrine secretion reaches the skin's surface, and the person's odourprint is significantly reduced. The frequency of the blocking allele is high in Korea, Japan, Mongolia, and northern China, with many people carrying two copies, and explains the low incidence of strong body odour in much of Asia. The blocking allele is rare in Africans and Caucasians, allowing skin bacteria access to the acids, with the consequential production of characteristically strong body smells in such people. About 2% of white Europeans carry two copies of the blocking allele, and lack discernible body odour. The

---

[68] Hummel, T., Krone, F., Lundström, J.N. *et al.* 2005 Androsadienone odor thresholds in adolescents. Hormones and Behavior 47: 306–310.

[69] Martin, A., Saathoff, M., Kuhn, F., *et al.* 2010 A functional ABCC11 allele is essential in the biochemical formation of human axillary odor. Journal of Investigative Dermatology 130: 529–540.

Huffington Post reports that prior to the engagement of pilots, Hainan Airlines conducts axillary 'sniff' tests and rejects applicants if they present with noticeable underarm odour.[70]

*   *   *

There's another important genetic effect determining human smell and it's wrapped up with the immune system. At first glance this may seem rather odd, since the immune system's role is to protect the body against disease, and smells could hardly be described as causing illness (though the Ancient Greeks thought otherwise). If you reflect on it for a moment, the apparent oddness of this isn't as great as it might first appear. The immune system exists to discriminate between self and non-self; invading organisms and foreign proteins threaten life through infection, and must be identified and mopped up with all speed. In a similar fashion, the sense of smell exists to enable a body to confront and interact with what is outside, to recognize what is different in others and take appropriate action. That action could be anything from eating the emitter, to running away from it, or to courting and mating with it. The set of genes that controls the development of the immune system is called the 'major histocompatibility complex', or MHC for short. Confusingly, in humans it's called the 'human leucocyte antigen', or HLA, but both terms describe the same bundle of genes doing the same job. The MHC contains many closely linked and highly variable genes. Humans have 140 MHC genes distributed in three gene families; a high number when compared with other mammals.

Kunio Yamazaki and others at the Monell Chemical Senses Center in Philadelphia demonstrated that mice could detect the MHC configuration of other mice by smelling their urine.[71] Females select male mates with MHC configurations which are a little different to their

---

[70] Huffington Post, Weird News 24th April 2013.
[71] Yamazaki, K., Boyse, E.A., Mike, V., *et al.* 1976 Control of mating preferences in mice by genes in the major histocompatibility complex. Journal of Experimental Medicine 144: 1324–1335.

own, and avoid mating with males with configurations similar to their own, perhaps to confer a degree of 'hybrid vigour' to better suit their offspring to resist infections. Other hypotheses to explain why mating preference might be for those with slightly different MHC carriers include the avoidance of inbreeding; mating with an unrelated individual will result in a widening of the gene stock for immunity, and with it comes improved health. The ability to assess MHC type by smell in mice is so strong that female mice can differentiate between males that are almost genetically identical to themselves, differing only with respect to their MHC genes. The MHC bundle of genes, in mice at least, is a powerful determinant of body smell, clearly influencing mate choice.

It's clear, again in mice, that the VNO is integral to the continuing conversation going on between the body's immune and nervous systems. In the Bruce effect, where the scent of a 'stranger' male perceived by a recently mated female is able to terminate her pregnancy, depends upon a functional VNO. If the female has had her VNO removed surgically prior to mating, the smell of a stranger no longer blocks her pregnancy. Through interrogating the immune system configuration of a potential sex partner *via* his or her body smell, the VNO provides genetically powerful information upon which the brain can make important behavioural decisions. As humans lack a VNO, are they denied information about the genetic configuration of potential life partners with whom to produce healthy offspring? We saw earlier that not all pheromone detection requires a VNO; in pigs, rabbits, and sexually experienced hamsters important behavioural decisions are based on messages from the nose alone. While little is known about how humans respond to the peptide molecules produced by the MHC genes, there is no evidence indicating that we can't perceive them, and some evidence that we can.

A convincing study of the relationship between MHC type and life-partner choice was carried out by Carole Ober, and her colleagues in Chicago, on MHC configurations in the Hutterite community of South Dakota.[72] About 400 Hutterites, a Caucasian religious group,

---

[72] Ober, C., Weitkamp, L.R., Cox, N., *et al.* 1997 HLA and mate choice in humans. American Journal of Human Genetics 61: 497–504.

emigrated to the USA in the 1870s following religious persecution in Europe and settled in three communal farms in South Dakota. Hutterites marry within their own communities and marriages are not parentally arranged. Young people are free to marry whomsoever they wish. Generally, Hutterites marry young, have large families, divorce is rare and adultery is strictly prohibited. Ober and her team looked at 411 marriages with a median marriage year of 1971 — marriages made long before modern media started to make its frequently unachievable demands on relationships. They took blood samples from all participants and identified five main types of MHC configuration within the community. The Hutterite community was an excellent choice for this type of study because the small initial population of 400, from which the present community has grown, carried a restricted number of MHC genes; far fewer than is found in the wider American population today. Hypothesizing that if MHC configuration had nothing to do with mate choice, the researchers would expect to find each MHC type randomly distributed across the marriages, as is assumed to be the case in outbred human populations. What they found, however, was a statistically significant reduction in the number of couples sharing an MHC type compared with couples of different MHC types, indicating that mate choice was far from random with respect to MHC genes. The significance of this study is that it was conducted on the observed frequencies of genes in the population and known life-pairings, and not on subjective ratings of how other people smell. Whether the scent of a putative partner played any role in courtship is unknown, and Ober and her colleagues make no claim for it; what the study showed was that a strong statistical relationship exists between MHC genes and mate choice.

It's long been known that the failure of pregnancy in couples with similar immune system configurations is higher than in couples with dissimilar types.[73] Doctors treating infertile couples find that the *in vitro* fertilization of eggs by sperm of MHC-similar couples is less successful than between MHC-dissimilar couples, and the same difficulty flows through to the implantation of embryos. This isn't a cut

---

[73] Ober, C., Elias, S., Kostyu, D.D. *et al.* 1992 Decreased fecundity in Hutterite couples sharing HLA-DR. American Journal of Human Genetics 50: 6–14.

and dried issue, because the great size of the MHC gene stock allows for an almost infinite range of similarities and dissimilarities between couples, and there's an unknown threshold of MHC similarity at which failure of pregnancy becomes a problem. Close examination of the data continues to indicate that fewer men and women with identical MHC-type partners have children than men and women with dissimilar MHC configurations. Attempts to demonstrate whether a person's body smell, reflecting their MHC-type, plays any part in his or her mate choice have not delivered clear-cut outcomes.[74] In large measure this is the result of differing methodologies used to obtain results about odour preferences in intimate relationships, but there's enough evidence implicating the MHC to suggest that further work be conducted. Humans aren't mice, but are complex creatures in which even the most fundamental biological decisions are made on a suite of criteria, conscious and subconscious, and employing the full range of senses.

*   *   *

So, what do we know about how humans respond to body odour; do we smell nicely, and do we like how other people smell? All things considered, the answers would have to be 'not really' and 'not very much', respectively, at least as far as casual, unperfumed meetings are concerned. Researchers into human body scent have developed a standard methodology for investigating these matters, in which underarm secretions are collected from clean cotton T-shirts worn overnight, and bagged and frozen first thing the next morning. An alternative methodology sometimes employed is for cotton pads to be taped into the armpit and worn overnight. Participants in such experiments are told that their bed linen should be washed before the collection period in scent-free washing powder. They shouldn't use scented soap, deodorants, perfumes, or aftershave, and they should use only unscented soap and shower products. They should refrain

---

[74] Havlicek, J., and Roberts, S.C. 2009 MHC-related mate choice in humans: a review. Psychoneuroendocrinology 34: 497–451.

from smoking, and from eating rich and spicy foods such as strong cheeses, asparagus, garlic, onions, green chili, pepperoni, cabbage, lamb, and celery, all of which influence body smell. They are asked not to share a bed with anyone during the duration of the study and, depending on the specific objective of the study, may be asked to abstain from sex. If an objective of the study is to determine if women perceive male odour differently at different times of the cycle, only women not taking contraceptive pills are allowed to participate. This is because oral contraceptives work by mimicking the hormonal state of pregnancy, and not the state of sexual availability. Sometimes blood, urine, or saliva samples are taken before and after odour rating, particularly when changes in hormone levels are being examined. In the best experimental designs, the participants are questioned after they have handed in their soiled T-shirts if they have broken any of the conditions, and if they have, they're excluded from the study.

Most studies require a panel of sniffers to rate the odours from the T-shirts or pads for 'pleasantness', or to pair two samples from one donor from among a number of samples. In different studies, sniffers are asked to ascribe a number of descriptors to T-shirts, including 'pleasant', 'unpleasant', 'not unpleasant', 'sexy', and so on. In this and the next two pages, I've put such descriptors in inverted commas, to make clear they are terms used by the experimenters. The results that emerge from subsequent statistical treatments are all too often hampered by small sample sizes, and lack the degree of objectivity possible when the subjects are laboratory mice. There's a tendency, too, for research in this field to be picked up by mainstream media and poorly reported, with unsupported conclusions frequently advanced without qualification. Just because people rate the smell of one particular type of odourized pad or T-shirt as more, or less, pleasant than another does not logically translate into conclusions about the role of body smell in human mate choice, even if laboratory experiments with mice, rats, and other animals show that it does.

Having got this caveat out of the way, speaking generally, people are not very good at identifying their own, or other people's soiled T-shirts. The low success rate is typical of T-shirt studies; in technical terms, the noise-to-signal ratio is rather high, indicating that the cues are not very strong and are obscured by all the other smelly, and not so smelly stuff around. Nevertheless, there have been enough studies carried out in laboratories around the world to enable some general conclusions to be teased out, and they're as follows.

Women are better at identifying odours than men and generally rate their own smell as 'pleasant'. Men generally rate their own smells as 'unpleasant'. The odour of the partners of both men and women is generally described as 'pleasant' or, at least, 'not unpleasant', and both men and women attach the word 'pleasant' more often to female smell than to male smell. Male smell is generally described by both sexes as being 'more intense' than that of women. Men generally rate the smell of women in the fertile phase of their cycle as 'more pleasant' and 'sexy' than the smell of women in the non-fertile phase of the menstrual cycle. Mothers can readily identify their own children's scents but not that of their stepchildren, and preadolescent children can identify their full siblings but not their half-siblings or step-siblings. Mothers have difficulty distinguishing between the scents of their identical twins, though they can recognize each of their twins as their own.

It's a fact that the axilla shows quite strong handedness. Right-handed people sweat more from their right axillae, while in left-handers the left axilla dominates. There's a suggestion that the dominant axilla produces more of the steroid derivative androstenone than the non-dominant axilla, though more work needs to be carried out to confirm this finding. For left-handed men, at least, there's also an intriguing observation that the left axilla produces a more intense and more masculine scent than the right axilla, when evaluated by women not taking oral contraceptives and in the fertile phase of their cycles.[75] A similar but opposite effect is not seen for right-handed men, however, presenting interpretational difficulties of the kind so often encountered in T-shirt evaluations. Nevertheless, there are a sufficient number of tantalizing leads that need to be followed up before a full understanding of axillary phenomena emerges.

---

[75] Ferdenzi, C., Schaal, B., and Roberts, S.C. 2009 Human axillary odor: are there side-related perceptual differences? Chemical Senses 34(7): 565–571.

Despite experimental and interpretational difficulties, the evidence points to a relationship between the configuration of a person's immune system and armpit secretion, just as there is in mouse urine. Women and men dislike the smell of T-shirts from donors with an immune configuration type similar to their own more than they dislike the smell of shirts of donors with different configurations. There's no difference between how men and women respond, more positively scoring shirts from donors with different MHC configurations to their own. Women sniffing male secretions from donors with MHC configurations similar to their own frequently report the smell as 'unpleasant'. The highest reported preference by women is for odours from men that have MHC configurations neither too similar, nor too dissimilar to their own. Both men and women are reminded of their current, or past partners, when sniffing a T-shirt from a donor whose MHC type was dissimilar to their own.

*   *   *

Beauty is an unlikely player in the smell story and it plays an interesting part. The Ancient Egyptians would only allow their (male) Pharaohs to marry beautiful women, because it was thought that if they did, they would more likely produce beautiful children. The beautiful were supposed to be the best rulers, thought to possess the highest skills of learning, ability to grasp new concepts, and make good judgments and wise decisions. *Callipedy,* as this process of begetting beautiful children was called, was based on the view that the fittest mates were the ones with greatest physical beauty, as revealed by the face and stature. Much research has gone into the analysis of human beauty and while there are a huge number of cultural elements influencing what makes a person beautiful, there are also a number of universals operating across all cultures. Central amongst these is symmetry. A symmetrical face is generally described as beautiful; the teeth are prime drivers of facial symmetry, something for which the orthodontic industry can be thankful. Skeletal symmetry leads to good body posture and an attractive gait, and so on.

Randy Thornhill at the University of New Mexico has conducted quite a lot of work on what is called 'fluctuating asymmetry'. This is asymmetry of anatomical features that's not necessarily caused by genetics, but by the slight developmental perturbations occurring in the womb before birth.[76] A number of environmental stressors can cause asymmetry, including whether the mother is carrying a high load of parasites and pathogens. High parasite and pathogen loads are commonly found in mammals, and were also, presumably, a scourge of our distant ancestors. Such unwelcome fellow travellers may help explain the unusually high number of MHC genes in humans. Thornhill and his colleagues defined symmetry rather more objectively that did the Ancient Egyptians by measuring physical features such as right and left ear length, ear width, elbow width, wrist width, ankle width, foot breadth, and the lengths of all fingers except the thumb; features that you might not immediately look for in a person you'd describe as beautiful. They recruited groups of men and women to use as odour donors, in a trial with male and female sniffers, to examine if the degree of anatomical symmetry played any part in body smell preference. All donors were measured for their fluctuating asymmetry characteristics, and sniffers had no knowledge of the sex, or the symmetry status of the owners of the T-shirts they were sniffing.

The results were astonishing. During the fertile phase of their cycles, women sniffers preferred the smell of more symmetrical men to the smell of less symmetrical men. Women on the contraceptive pill, and not displaying normal oestrous cycles, showed no preference between the two. It's easy to over-interpret these results, and until more replications are conducted in other laboratories under identical conditions we should approach them with caution. However, they invite us to consider that at least one aspect of a person's beauty — their body's symmetry perhaps standing as a proxy for their health status — may be coded in their individual scent signatures.

*   *   *

---

[76] Thornhill, R., Gangestad, S.W., Miller, R., et al. 2003 Major histocompatibility complex genes, symmetry and body scent attractiveness in men and women. Behavioral Ecology 14(5): 668–678.

There's plenty of evidence that smell plays a part in human kin recognition, just as it does in animals.[77] Newborn infants become imprinted on their mothers' breast smell, and mothers quickly recognize their own newborns from their smell alone, as also happens in mice, rabbits, and sheep, and probably most other mammals. Sniffers unrelated to mothers and their children are able to pair T-shirts worn by mothers and their eight-year-old children by smell alone, picking up a family resemblance in scent profile. If MHC and olfactory genes are linked, this is to be expected.

Identical and non-identical twins offer opportunities to look at various aspects of the genetics of body smell. Early studies used tracker dogs to discriminate between identical twins, with limited results probably due to the strongly confounding influence of environmental factors affecting body odour. Dogs can discriminate between twins if they differ either in genetic constitution (i.e., they are non-identical) or in environmental factors (i.e., they are identical but have been brought up in different households), but not in both. Craig Roberts and a team from the University of Liverpool have conducted one of the most careful studies on the similarity of twin body odour using humans as the discriminators, rather than dogs.[78]

For odour donors, Roberts' team used groups of identical and non-identical twins who did not cohabit but who lived in different households and were exposed to different environmental factors. The discriminators were a group of over a hundred members of the public associated with a university medical school. The standard experimental T-shirt methodology was employed. The hypothesis behind the study was that if there was no genetic component to body odour, sniffers would match identical twins' smells no better than they would match non-identical twins' smells. In fact, the sniffers matched the body smells of identical twins better than would be expected by chance, but not statistically significantly better, and the odours of non-identical twins as no better than by chance. These results don't confirm the idea that there's a genetic component to body odour

---

[77] Porter, R.H. 1999 Olfaction and human kin recognition. Genetica 104: 259–263.
[78] Roberts, S.C., Gosling, M.L., Spector, T.D., *et al.* 2005 Body odour similarity in non-cohabiting twins. Chemical Senses 30: 651–656.

quality that's not overridden by environmental factors, but neither do they quash it. The results revealed what all scent-sniffing studies show, namely that cues contained in body smells are subtle. Scent cues are not like the visual signals that are the distinguishing hallmarks of a person's identity, instantly recognized and immediately apparent.

As humans grow old, their body odour changes in many ways that are popularly reported as being 'unpleasant'. The characteristic smell of old peoples' homes may not reflect so much a change in body odour of the residents as the smell of stale urine, intestinal problems, oral hygiene, and the like. In trials on underarm odour taken from old, middle aged, and young people, the 'least unpleasant' axillary odours were produced by the elderly, and the 'most unpleasant' by middle-aged men. Teams of young and middle-aged sniffers asked to pick out which of a series of glass jars contained axillary pads from old people, with the other jars containing pads from young and middle-aged people, most easily selected the jars with elderly people's scent. It seems the odorous social stigmata associated with age is not caused by changes in body odour and is unjustified by rational analysis.

It's often held there is a smell associated with fear. Dogs and other animals are supposed to be able to detect its aura as soon as they get a whiff of human scent. There's some evidence that there is a smell associated with fear, or at least, associated with stress, and that it's expressed in underarm sweat, though the signals are hardly overwhelming. In an investigation of whether odorous stress cues affect neural activity related to the evaluation of emotional stimuli, sweat from first-time skydivers was presented to subjects who were asked to look at pictures of human faces showing a range of expressions.[79] As a control smell, pads taken from donors who had been running on an exercise treadmill were used. Volunteer sniffers, acting as fear detectors, were fitted with skullcaps designed to measure the brain's patterns of electrical activity, with electrodes positioned over key centres of the brain. The results showed that when presented with the sweat from the skydivers, subjects displayed a pattern of

---

[79] Rubin, D., Botanov, Y., Hajcak, G., *et al.* 2012 Second-hand stress: inhalation of stress sweat enhances neural response to neutral faces. SCAN 7: 208–212.

brain activity that is normally seen when sustained attention is paid to a stimulus. This effect wasn't seen when they were presented with exercise sweat. It was as if odour cues produced by a stressed person enhanced the processing of it by perceivers, as though it warranted extra attention. The physiological mechanism behind this interpretation is obscure. We can rule out a major effect of apocrine secretion because we know it takes several hours for the initially odourless material to take on its characteristic smell. Most likely, the cues will be found to be contained in the true cooling sweat, copiously produced by the armpit during stressful situations, or possibly apocrine secretion carries more than just a single set of messages. Wherever the cues originate, preliminary work such as this suggests that body smell can convey emotional status, in addition to sexual orientation, gender, immune system configuration, and sexual condition.

*   *   *

The possibility that humans might produce and receive sex pheromones, and that they play a part in human sexual behaviour, has been debated for decades. The first indication that humans might produce scents that could interfere with the sexual biology of others came over 40 years ago, when a psychology student at Wellesley College in Boston named Martha McClintock became interested in old wives tales about the cycles of women who live together becoming synchronized.[80] She asked all 135 women sharing her dormitory block to record the start of their menstrual periods during the college year, and concluded that the mean difference in period start dates at the beginning of the year was reduced as the year went on, thus indicating, she concluded, a trend towards menstrual cycle synchrony. Her paper, published in the prestigious journal *Nature*, caused a flurry of interest and intrigue that spawned a host of studies designed to confirm her ideas. Just as many supported the menstrual synchrony hypothesis as rejected it.

The problem with this sort of research is that conclusions depend upon statistical analyses of the data about phenomena that are

[80] McClintock, M. 1971 Menstrual synchrony and suppression. Nature 229: 244–249.

inherently variable. Although the mean cycle length of the human menstrual cycle is about 28 days, cycles of between 21 and 35 days are not uncommon. Women with longer and shorter cycles will invariably find that in some months their times of menstruation will overlap, and female-produced pheromones are not required to explain it. If the human sexual cycle was manipulated by pheromones, as happens in mice and lots of other animals, unequivocal proof would have been forthcoming by now. A further problem is that no mechanism has been identified, other than the VNO-based pheromonal mechanism that occurs in mice, which, as we know, doesn't exist in humans.

The question those who support the idea of menstrual synchrony must answer is what evolutionary advantage would accrue from it? To have all females simultaneously in breeding condition is a prereq-uisite for seasonally breeding species, such as red deer, elephant seals, or sage grouse, but brings no advantage to species evolved to live monogamously. We must conclude that no unequivocal case has been made for pheromonally induced menstrual synchrony in humans.

Almost three decades after her first foray into human olfaction, McClintock found that underarm secretions from women in the fertile phase of the menstrual cycle, when wiped on the upper lip of women who were about to ovulate, accelerated a surge of luteinizing hormone, thus shortening the recipients' cycles.[81] She found the reverse also to be true, namely that secretions from women in the non-fertile phase delayed the surge in luteinizing hormone in recipients, so lengthening their cycles. She paid proper attention to controls and determined the phases of the cycle that donors and recipients were in by assess-ments of luteinizing hormone from daily urine samples, a reliable indicator of whether a woman is in the follicular or ovulatory phase of her cycle.

*   *   *

The best place to look for the influence of pheromones produced by the opposite sex would be in the hypothalamus of the brain, sitting as

---

[81] Stern, K., and McClintock, M.K. 1998 Regulation of ovulation by human phero-mones. Nature 392: 177–179.

it does at the start of the BPG chain. Technically this presents problems, but there are excellent proxy hormones that are easily measured. The release of GnRH from the hypothalamus stimulates the pituitary to produce and release LH and FSH into the blood. These hormones are later excreted and can be easily assessed from urine samples.

Studies on the effects of male body odour on the measurable levels of sex hormones in women and body odour of women on sex hormones in men, offer promising grounds for analysis. A carefully managed study undertaken at the Monell Chemical Senses Center by George Preti showed that the pulses of LH production in women speeded up by about 20%, following the application of an extract of male axillary organ secretion to the upper lip of a test subject.[82] Subjects did not know they were being used in a pheromone study and reported they could smell only a delicate, slightly alcoholic fragrance from the application — axillary extracts were treated with the same mild perfume used to scent a control alcohol. A reduction in latency, until the next pulse released by the pituitary, came about during the first and second two-hourly applications of male odour, but didn't change thereafter. In this study, there were no reported changes in cycle length, as is seen in mice and sheep, but any stimulus that interferes with the production of LH has the potential to interfere with cycle length.

During this study the women tested were asked to fill in questionnaires one hour, and four hours into their testing periods to best describe their mood in terms of the descriptors: 'energetic', 'sensuous', 'tense', 'tired', 'calm', 'sexy', 'anxious', 'fatigued' and 'active'. They could answer anything from 'I am not at all ...' to 'I am extremely ...' to the descriptors, and the statistical interpretation of the results was as robust as these things can be. It turned out that the application of male smell reduced feelings of tensions, and increased feelings of calmness and lack of anxiety. No other descriptors showed any change.

The effect of human odours on mood has been examined by a few researchers, all using slightly different methodologies and sample

---

[82] Preti, G., Wysocki, C.J., Barnhart, K.T., *et al.* 2003 Male axillary extracts contain pheromones that affect pulsatile secretion of luteinizing hormone and mood in women recipients. Biology of Reproduction 68: 2107–2113.

sizes, and all asking slightly different questions. As human behaviour is so incredibly complicated it's not surprising that clear-cut, repeatable results have not been obtained. Martha McClintock's more recent work demonstrates that androstadienone can be regarded as the only known example of what she calls a mood 'modulator', though she stops short of calling it a pheromone.[83] Through carefully designed studies employing men and women responding to underarm extracts of both sexes placed on the upper lip, she showed that treated subjects engage stimuli with a higher level of emotional significance than untreated control subjects. The extract was accurately diluted such that only a minute and repeatable amount of the steroid was applied, that was well below the concentration at which any subject could say she or he could smell it. As a precaution, oil of cloves was added to provide an equivalent fragrance for experimental and control applications alike to mask any possible smell of the steroid. Men and women responded alike to computerized tasks, with no indication that the mood of either sex was more affected than the other. McClintock and her colleagues concluded that androstadienone modulates the psychological state of the receiver by inducing the brain to concentrate more closely on emotional information which, the experimenters concluded, could conceivably influence subsequent judgments and actions.

It's not known why men's armpits produce more androstadienone than women's; it may simply be because men have more testosterone derivatives in their blood (which they do), or more of the coryneform bacteria that break down its biochemical precursor than do women, but it may be because of something else. At the time axillary secretion is produced by the apocrine glands, there is no difference between men and women — the signature differences appear only after the bacteria have had time to do their work. The idea that androstadienone is a male pheromone that attracts women towards men isn't borne out by any experimental studies, and is based more on what happens in pigs than in humans. Modulator smell signals don't signal anything about the producer, such as her/his sexual

---

[83] Hummer, T., and McClintock, M.K. 2009 Putative human pheromone androstadienone attunes the mind specifically to emotional information. Hormones and Behavior 55: 548–559.

status, physical condition, symmetry status, social dominance, or anything else. Those things are the province of signaller pheromones and to date no signalling pheromone has been confirmed in humans.

Androstadienone has no effect on cognitive tasks and doesn't release any stereotyped behaviour, though there's evidence that its smell raises the blood levels of the stress hormone, called 'cortisol', in women. Claire Wyart and her team at Berkeley, California, showed that about 15 minutes after sniffing pure androstadienone, cortisol levels in women rise sharply, and stay that way for up to an hour.[84] The researchers wired-up their subjects so they could measure a number of physiological parameters such as heart rate, blood pressure, skin temperature and conductance, and so on. They also assessed subjects' mood using a well-worked mood scale that was applied after the subjects had watched humorous, sad, or erotic videos. Cortisol levels were measured from saliva samples taken before the experiment began, after exposure to the smell of androstadienone, and again after each mood video. The results were statistically quite clear-cut. Exposure to androstadienone increased physiological arousal, generated a more positive mood, and increased reported sexual arousal. The male axillary secretion contains much more than only androstadienone however, and Wyart's experiment doesn't say if the physiological and mood changes she observed are in response to the entire bouquet, or to only a part of it.

Androstadienone occurs in all male mammals, and there's some evidence it brings about a rise in stress hormones across a range of species. A recently developed, and very sensitive pain-measuring scale in mice, based on the facial grimaces the rodents make as an expression of stress, has enabled researchers to examine the phenomenon. Not only does it appear that experimental mice are responsive to the sex of their handlers, but it also raises a potentially important issue for medical research.[85] Stress suppresses pain — a more than

[84] Wyart, C., Webster, W.W., Chen, J.H., *et al.* 2007 Smelling a single component of male sweat alters levels of cortisol in women. The Journal of Neuroscience 27(6): 1261–1265.

[85] Mogil J.S., and twenty others 2014 Olfactory exposure to males, including men, cause stress and related analgesia in rodents. Nature Methods, doi: 10.1038/NMETH.2935.

useful outcome if an animal is attacked by a predator in the wild, but manages to escape. Mice handled by male experimenters, but not by female experimenters, show an elevated level of stress hormone cortisol, and a suppression of pain as revealed by the pain scale. If a female researcher places on the bench a man's worn T-shirt, the effect is as strong as when a male researcher handles the mice. Bedding from male guinea pigs, male dogs and male cats similarly brings about suppression of pain, but bedding from castrated males of the same species is ineffective, indicating that the effect depends upon the presence of testosterone. The sex of experimenters is seldom — if ever — recorded in published reports of medical breakthroughs; instances in which other laboratories are unable to reproduce the results of the one announcing the breakthrough may be because handler-induced stress was a factor in one laboratory, but not in another. The matter requires careful attention.

As men don't have an oestrous cycle, any effect of female odour on their sexual biology should be looked for in the target organs of the sex hormones released by the pituitary. Pituitary LH in men stimulates the testes to produce testosterone, a hormone easily assayed by a saliva test. Fewer studies on the effect of female odour on males than *vice versa* have been conducted, so the emerging story isn't as well understood, but men exposed to the body smell of women (overnight T-shirt technology once again, usual limitations on diet, soap, oral contraceptives, sexual partners etc.) show perturbations in their testosterone levels. Testosterone is depressed by the body odour of women during the non-fertile phase of their cycles but not depressed during the fertile phase, indicating that men's noses are sending signals to their brains about the stage of oestrus of the woman whose armpit scent they are perceiving, though whether the signals change male behaviour isn't known. Testosterone is the hormone that controls mating behaviour and other traits that are associated with maleness, including *inter alia*, risk-taking behaviour, social dominance and aggression, and its levels in men respond to many stimuli including social interactions with women, participation in competitive environments such as on the sports field, and watching arousing films.

*   *   *

By way of a historical footnote, it's interesting to note that when King David lay close to death, his servants thought the proximity of a nubile young woman might restore him to health. The Bible (*New Revised Standard Version* 1989) starts the Book of Kings this way:

> 'King David was old and advanced in years; and although they covered him with clothes, he could not get warm. So his servants said to him, "Let a young virgin be sought for my lord the king, and let her wait on the king, and be his attendant; let her lie in your bosom, so that my lord the king may be warm." So they searched for a beautiful girl throughout all the territory of Israel, and found Abishag the Shunammite, and brought her to the king. The girl was very beautiful. She became the king's attendant and served him, but the king did not know her sexually.'

Based on our modern understanding of the effects of female body odour on males, it's at least possible that a short-lived restorative effect could come about through an odour-induced effect on the old man's testosterone levels, and with it a general feeling of well-being.

The restorative practice became known as 'shunamitism', after Abishag's home town, and was much favoured by physicians during the 17th and 18th centuries, and doubtless also by their patients! Biblical scholars note that Abishag eventually married, or became a concubine of King Solomon (he had 700 wives and 300 concubines), and may have been the subject for the *Song of Solomon*.

It's worth noting that a recent study conducted in Japan on elderly men during a regular exercise class found that exposure to a putative female pheromone brought about an increase in the level of testosterone in the blood; subjects exposed to a placebo smell showed no such elevation. Although the study was preliminary and ran for only a few weeks, the possibility of a pheromone-induced elevation of testosterone in men could have significant health outcomes, including reversal of age-related conditions, such as reduction in muscle mass. 'Shunamitism' may yet be found to be a viable therapy, though medical insurance companies might take a bit of convincing!

*   *   *

A final — if slightly uncomfortable — thought about your body's smell. A consequence of walking upright, with your nostrils directed downwards rather than to the front as in dogs, is that the air reaching them has flowed upwards and over your body from your feet, driven by convection. Not only does it carry the smell of your many scent glands and of your sweat, it also contains millions of sloughed skin cells that travel up your body in the layer of convected air lying next to the skin, along with dust, pollen, spores, and the like. Like the thought of it or not, every breath you take is laden with this particulate broth. Most of it escapes your nose and carries on upwards, to be vented off your shoulders and the top of your head. If you walk at 1.3 metres per second — about 3 mph — you leave behind some 500 skin flakes every metre, and more if you move faster. You shed about 1 million flakes every minute, and breathe in between 6,000 and 50,000 in every breath! As you walk, an odour plume develops behind you, characterized by eddies and vortices, the exact location and duration of which depends upon local wind conditions (Figure 7.2). Next time your head's turned by the scent of a person walking by, remember that you are sampling as much their sloughed skin cells as their perfume!

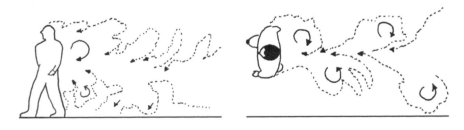

Figure 7.2

Sketch of the vortices and eddies created in the air, by a walking person. Side view on the left; top view on the right. Redrawn from Settles G.S., 2005, Journal of Fluids Engineering, vol 127, page 209.

Chapter 8

Armpit Evolution

In 1880, the French novelist Joris-Karl Huysmans shocked the genteel citizens of Paris with a little vignette entitled *Le Gousset* (*The Armpit*), which he wrote as one of a series of sketches about that city's inhabitants.[86] The scent of the armpits of country women conveyed, he declared, 'something of the relish of wild duck cooked with olives and the sharp odour of shallot', while in the grand ballrooms of Paris 'the aroma is of ammoniated valerian, of chlorinated urine, brutally accentuated sometimes, even with a slight scent of prussic acid about it, a faint whiff of overripe peaches'. In brunettes and dark-haired women, he judged, it was 'audacious and sometimes fatiguing'; in redheads it was 'sharp and fierce'; and in blondes 'the armpit is heady as some sugared wines'. By probing where few before him had dared to go, his critics hinted he'd doomed his work to be placed behind the desk in public libraries, away from innocent eyes. Twenty years before the *fin-de-siècle* public sensibility in Paris wasn't ready yet to be educated about human armpits, despite the popularity of *risqué* Bohemian lifestyles in that, then, most liberated of European cities.

Two millennia before Huysmans, the Roman poet Gaius Valerius Catullus offered advice to his friend Rufus, who had complained that he was having little success with women. Catullus, who had heard a rumour that Rufus kept a goat in his armpits, advised him that the least beauty imaginable would never bed with such a rank-smelling

---

[86] *Croquis Parisiennes* 1880 Henri Vaton, Libraire-Éditeur, Paris.

beast, and he should go and find some soap and water to fix the problem, or else stop complaining.

Here, then, are the two sides of the armpit story; axillary smells exalting writers to admiration bordering on worship on the one hand, while making intimate relations impossible with someone whose hygiene needs attending to, on the other. Why do Adam and Eve carry either ripe peaches or goats in their armpits?

* * *

No primates other than humans, chimpanzees, and gorillas have musk organs in their armpits, though many have a patch of scent gland tissue lying over the mid-line of the chest. Primatologists believe that ancestral primates had a broad band of scent-producing tissue stretching from one axilla across the chest to the other, and that one line of evolution saw the retention of a mid-chest gland, while the other line — the one we are in, retained the two extremities and lost the central patch. Gibbons belong to the other line and have well-developed mid-chest musk glands lying over the sternum. Gibbons' glands are structurally similar to human axillary glands, and like humans, both sexes have them. In these much-studied creatures, the glands have never been observed being used in any kind of behaviour in which scent from the gland is applied to another gibbon, or anything else. Exactly the same's true for humans, gorillas, and chimpanzees.

The human axillary scent gland consists of a patch of thickened hair-covered skin lying across the crease between the arm and the side of the chest, measuring about 50 mm long, 25 mm wide and up to 5 mm deep (Figure 8.1). It's packed with apocrine glands, together with many true sweat and sebaceous glands. Some of the individual apocrine glands are up to 2 mm in diameter, particularly those in the centre of the organ where the tissue is deepest. This description is true for Caucasians, Africans, Indians, and some others, but it's not true for Asians, who mostly have weakly developed axillary organs, along with the gene that prevents apocrine secretion from being carried to the surface. Only 2–3% of Asians have strong axillary smell, possibly reflecting minor past infusions of genes from elsewhere. Apocrine glands in Asian people are small and comfortably distributed through-out the axilla, not showing the dense packing characteristic of

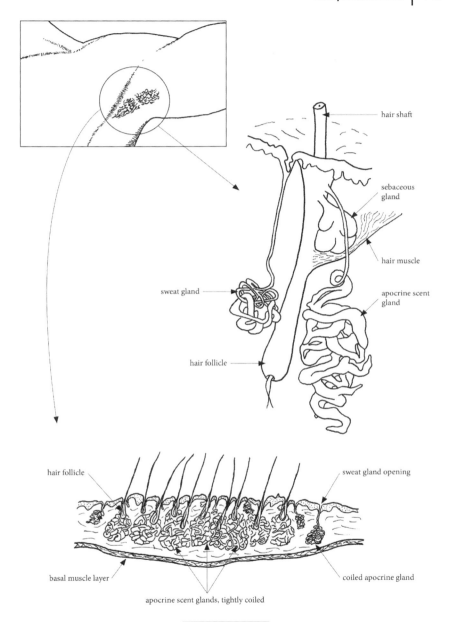

Figure 8.1

The human armpit (axilla), together with a single hair and its associated apocrine, sweat and sebaceous glands, and a section cut through the axilla, showing densely packed apocrine glands.

non-Asians. In half the population of Korea apocrine glands in the axillary organ are so sparse as to be almost absent.

The axillary organ is designed to do one thing and one thing only, and that is to produce scent. Development starts fairly late in embryonic life, at about the fifth or sixth month of pregnancy, but it's fully formed at birth. The apocrine glands within it remain dormant until puberty, when they start to develop rapidly under the influence of sex hormones. Under their influence the tissue patch deepens and thickens, and long springy hairs grow from deep follicles. There is little difference between the organs of men and women; other than the number of apocrine glands packed into women's organs is higher than in men's organs. There are no changes to the axillary organ during pregnancy or lactation, nor during the phases of the menstrual cycle. Axillary organs keep up their work for the rest of a person's life, waning only in very advanced age.

Apocrine glands release a viscous, milky, or slightly yellowish fluid. The trace of colour comes from pigment granules inside the glands that are expelled with the secretion. Occasionally the pigment granules become very dense, staining the secretion red, brown, or even greenish-blue. This distressing condition is known as 'chromidrosis' and gives rise to the common expression 'to sweat blood'. Hippopotamuses are said to sweat blood; apocrine glands all over the hippo's body produce a sweat containing reddish pigments that absorb ultraviolet light to protect the animal's delicate skin from sunburn, giving the appearance that they sweat blood. Chromidrosis in humans has nothing to do with sunlight — the armpits, arguably, see less sunlight than other parts of the body. The only truly effective treatment for chromidrosis is the surgical removal of the axillary organ. Excessive armpit sweating, called 'hyperhidrosis', is not uncommon during stressful encounters and can be eased with the application of deodorants containing astringent compounds, such as aluminium salts or lemon juice.

Apocrine glands release their secretion in response to adrenalin being released into the blood, and start producing fluid after the body has been stressed, such as happens when a person is frightened or attacked. Adrenalin is the hormone responsible for the 'flight or fight' response and is released into the blood by the adrenal glands the moment they receive the right signals from the hypothalamus *via* the pituitary. What adrenalin does is to prepare the body for action by increasing the heart rate, shunting blood away from the digestive system and towards the muscles and the brain, increasing an individual's ability to respond to a rapidly changing external environment. Sex is another stressor that releases adrenalin, but lest it be thought that poor Rufus' problem was that whenever he got near a woman he started to stink like a goat, we must remind ourselves that the apocrine secretion is initially odorless, taking on its characteristic aroma only after skin-dwelling bacteria have degraded it.

We might be tempted to think of the axillary organs as sexual characters, to be considered alongside the development of breasts and wide hips in women, and facial hair and deep voices in men — characteristics that develop under the control of oestrogen and testosterone. Where axillary organs differ is that while they need oestrogen and testosterone to switch them on, they are *present in both sexes*, and are *equally well developed in both men and women*. If they don't indicate a person's sex, what do they do?

*　　*　　*

To explain why armpits evolved, I need you to understand how animals choose their mates. In his book *On the Origin of Species*, Charles Darwin wrote of a phenomenon he called 'sexual selection', defining this branch of the broader concept of natural selection as depending '...not on a struggle for existence, but on a struggle between the males for possession of the females; the result is not death to the unsuccessful competitor, but few or no offspring.' He noted that the differences in the appearance of the sexes has been brought about by sexual selection, and mentioned the beautiful tail fans of male peacocks, the antlers of male deer, and the gaudy plumage of male birds of paradise as examples of showy advertisements used by males to attract females. The point about sexual selection is that not all individuals in a population of animals will leave the same number of offspring; there will be competition within the sexes for the right to mate. In the vast majority of species it's the males that compete amongst themselves for the right to mate, and thus they that carry fine adornments of feather, horn and antler, song and smell. As with just about everything in biology, there are rare exceptions in which it's the females who compete among themselves for the right to breed — the suppression of reproductive development in female African mole rats by a single queen is a rare mammalian example, though there are many examples to be found in the social insects (ants, bees, termites etc.).

In the mammals, investment in the next generation by males and females is anything but equal. Males produce sperm and deliver them into the female. That's it. When that's done, their biological role in handing on their genes to the next generation is over. For females, however, the real job is just about to start. After the eggs have been fertilized, embryos have to be carried through a long developmental period, punctuated with the infant's birth and subsequent lactation. All during this time the female has to find more food than usual, because she's growing one or more fetuses. In some species, such as antelopes, horses, whales, and guinea pigs, the young are born in an advanced state, and able to run with the herd, or swim unaided within a few minutes of birth. But humans are born helpless, in need of constant care and assistance until they are sufficiently developed to fend for themselves.

If *mating* is about the processes leading to the fusion of a sperm and an egg, *reproduction* is all about leaving one's genes to the next generation. The idea that genes are what matter in evolution was popularized by the Oxford biologist Richard Dawkins, who wrote about how genes drive animal structure and behaviour in his book *The Selfish Gene*.[87] The basis of the selfish-gene argument is that the bodies carrying genes and the behaviours they display are but slaves to the genes, and once the body's job is done, it's expendable. Sexual selection has moulded bodies and biology to best fit them for ensuring the genes they are carrying are transferred into the next generation, a process that might require the attendance of both father and mother for quite some time.

---

[87]Dawkins, R. 1976 *The Selfish Gene*. Oxford University Press, Oxford.

Across the whole animal kingdom, there are countless ways in which male and female animals serve their genes in the name of reproduction, both physically and behaviourally. We've seen that mating in some invertebrates may be a vague activity, in which eggs and sperm are cast into the ocean in the hopes that they will find one another and perpetuate the species. For species in which the female carries fertilized eggs within her body, physical and behavioural systems have evolved for males to deposit their sperm inside their bodies. Often there are bizarre consequences to the close encounters necessary for internal fertilization, such as when female spiders and praying mantises eat their mates immediately after mating. As weird as this may sound, if you realize that food is always in short supply in nature, it makes sense that once the male has done his fertilization job, he can further the chances that his genes will survive by providing some nutriment for his mate who will, as a result, be better able to look after his young. The trick here is to stop looking at the world through human eyes and try to see it from a gene's perspective.

*   *   *

Over 40 years ago the British zoologist Desmond Morris examined the many features of human anatomy, behaviour, and physiology that evolved to reinforce the pair-bond in what he described as 'a highly-sexed, pair-bonding species with many unique features; a complicated blend of primate ancestry with extensive carnivore modifications.' His book, *The Naked Ape*, is well worth reading if you haven't already done so.[88] He argues that the human body is endowed with a huge suite of features, evolved to make sex more rewarding and secure, for the purpose of keeping sexual interest going between parents *for as long as necessary for their young to become independent.*

Morris argued that humans evolved features that release sexual interest through stimulating all the senses, though the visual and tactile senses predominate. He placed great importance on the frontalization

---

[88] Morris, D. 1967 *The Naked Ape. A Zoologist's Study of the Human Animal.* Jonathan Cape, London.

of sexual signalling. Arguing that 'face-to-face sex is personalized sex', he showed how visual and tactile releasers, such as large breasts with distinctive areolae on the upper front of the body; lips that expose part of the sensitive pink mouth lining; hairless skin richly endowed with nerve endings particularly around the genitals, nipples, and earlobes; sensitive hands evolved from complex hunting; luxuriant pubic, axillary, and facial hair; and neck and chest blushing, all evolved to make, in his words, 'sex sexier'. He noted, further, that human males have by far the largest penises of any of the great apes, resulting in copulation providing mutual rewards of a kind unparalleled elsewhere in the animal kingdom.

Morris drew attention to the fact that once the sexual characteristics have developed at puberty, they remain present throughout life. The usual situation in nature is for sex-attraction or pair-bonding adornments to be shut down each year after the breeding season is over. Birds moult their plumage, deer shed their antlers, frogs stop croaking, and male alligators cease emitting the scent of roses once the breeding period has passed (yes, male alligators smell of old-fashioned rose gardens during the breeding season!). Where humans differ from most animals is that they are sexually active from puberty onwards, and their breeding is not seasonally restricted. Females are sexually receptive at all stages of the oestrous cycle and males are sexually primed from puberty onwards. Morris stresses the importance of continuous availability of sex to a species that has evolved from a leaf-eating primate living in small family-based groups to a species that now lives communally for the benefits that cooperative big-game hunting can give it. Humans are, he notes, the sexiest of all the great apes, only challenged in the field of promiscuity by the bonobo, a species of chimpanzee that uses sex as currency in just about all its social interactions.

*   *   *

What does the study of humankind tell us about the manipulative power of sexual selection? The first point to note is that our off-spring are born after a long gestation, and develop slowly through infancy and childhood taking some years before they are able to fend for themselves, resulting in our species having a low lifetime reproductive output. If you go back to the time before there were handy shops to supply our nutritional needs, both parents had to hunt and gather food, if their family was to prosper. As our ancestors evolved, the transference of knowledge and technological skills to the young became increasingly important, requiring further lengthy parental input.

The second point is that all the evidence points to our ancestors having lived monogamously. There are some differences in size between the sexes, but they are not dramatic. Men are 4% taller and 8% heavier than women, and can run 11% faster. The fact these differences exist at all suggests that, sometime in our past, our ancestors competed for multiple mates — a situation that we've seen in gorillas and orangutans, which favours significant sexual dimorphism. Study of fossilized skeletons of our 3-million-year old ancestor 'Lucy', *Australopithecus afarensis*, shows a degree of sexual dimorphism about the same as we see in modern humans, suggesting Lucy's peers practiced monogamy to about the same extent as we do. Because the degree of difference between the sexes isn't nearly as great as in polygynous mammals, we must conclude that the evolutionary pressure for males to monopolize multiple mates wasn't very strong.

The final point is that we are endowed with a scent-signalling device that's clearly related to sexual development, equally well developed in men and women, and not used in any behavioural scent-marking context. It perfectly fits the bill as an adornment evolved for the mutual benefit of the members of a reproductive partnership.

Putting all the anatomical and comparative data together, we can conclude that our ancestors evolved from a mating system that encouraged one male to remain with its mate until such time as the offspring was independent. In other words, monogamy was the most likely type of mating arrangement practiced by our ancestors. Before addressing the all-important question of how do males and females of monogamous species choose one another, I should pause to clarify what sort of mating arrangement pertains to modern humans. A clarification is needed, because already some readers will be saying the emerging argument is academic, idealized, and doesn't square with what can be read about daily in the gossip magazines.

*   *   *

There are plenty of examples from anthropological research into the social biology of remote tribes and wider human societies in which monogamy is or isn't always practiced. Among Aka pygmies of the Central African Republic, for example, men with brothers in their camp marry many wives and play little part in rearing their own sometimes quite numerous children, while men without brothers marry only one woman and contribute much to their children's upbringing. Even within this one tribe it's clear that one size clearly doesn't fit all. In wider society there are numerous examples of children being brought up by expanded family members or by the whole community itself, indicating that any biological argument that defines monogamy as the only arrangement for humans is flawed.

I don't deny that 21st century humankind has organized its societies differently from the way in which natural selection arranged them, and neither do I hold that biology is infallible. Evolution is like an immense experiment, with all manner of genetic variations being tried out in an external environment that is itself constantly changing. A few adaptations will survive, but most will not. If trying and testing didn't happen, there would be no adaptation to a changing environment, something that the fossil record shows us has been happening from the very beginning of life on Earth. As Matt Ridley points out, it's simply not possible to say categorically that the mating patterns of modern humans can be categorized as monogamous, polygamous, polyandrous, or anything else for that matter.[89] Throughout history there have been examples of societies showing all possible types of mating arrangements and probably many more than you could think of. All manner of geographical, nutritional, and population-size issues combine to determine what system is best for one society, at any one time, living under a particular set of environmental conditions — something

---

[89] Ridley, M. 1993 *The Red Queen. Sex and the Evolution of Human Nature.* Harper Collins, London.

exploited by the Aka with such skill. If you consider that our species has a history going back only a quarter of a million years or so, while that of most other species with which we compare ourselves with go back several millions of years, it's unrealistic to expect that our corner of the huge evolutionary experiment is painted yet in its final colours.

Readers unconvinced by Darwin's theory of evolution by natural selection might say the monogamy argument is too narrow, and that the concept of the single male family as the basis for human evolution is only one among many. That's as may be, and I'll return to issues of alternative family structures and the smell environment later, but the wide and comparative study of animal biology leads to only one conclusion about the mating arrangements of our primate ancestors; they lived mainly in single male families and males participated in rearing the young. Modern societies, based more on economic than ecological principles, are very different from the hand-to-mouth principles that governed our ancestors' evolution, leaving the way open for some societies to pursue alternative mating strategies. The fact that biological analysis doesn't include every single instance of how humans organize their societies doesn't invalidate the general principle. If polygyny, or some other system of multiple mate taking had been more commonplace in our ancestry, the physical difference between men and women would be more like that seen in gorillas or orang-utans than it is.

*   *   *

Charles Darwin noted that in monogamous mating relationships, members of a putative pair must choose one another *mutually*, as a large slice of their lifetimes will be invested in the product of perhaps a single mating. Subconsciously they must be able to assess the genetic quality and health status of their mates. In common parlance, males must be choosy and females picky. He observed:[90]

'There are, however, many animals in which the sexes resemble each other, both being furnished with the same ornaments, which

---

[90]Darwin, C. 1871 *The Descent of Man, and Selection in Relation to Sex*. John Murray, London.

analogy would lead us to attribute to the agency of sexual selection. In such cases it may be suggested with more plausibility, that there has been a double or mutual process of sexual selection; the more vigorous and precocious females selecting the more attractive and vigorous males, the latter rejecting all except the more attractive females.'

Ornaments carried by *both* sexes are far less likely to play a role in the *initial* attraction by one sex or the other than a flamboyant tail fan or other flashy advertisement carried by just one sex. Evidence that mutual ornaments contribute to the continuous reinforcement of the pair-bond and consequential reproductive benefits isn't easy to come by, but Maria Servedio and her colleagues at the University of North Carolina have looked at it in birds. She subjected a large amount of data on bird species (recall that monogamy is common in birds) exhibiting adornments carried by both sexes, such as colour patches, and feathery tufts and crests, to a theoretical analysis of the likely fixation of genes for mutual adornment under scenarios including their use in mate attraction *versus* stimulation of a partner to invest more in his or her brood.[91] Her modelling shows that genes for mutual adornments elevate parental investment over mate attraction, and have their effect on the survival of young through maintaining parental cooperation until the young are independent. In other words, there's evidence that mutual adornments supports monogamy and doesn't support mating systems in which one sex seeks out multiple mating opportunities.

Research has been conducted on how ornaments are related to an individual's genetic quality and health status, and again the best evidence comes from studies on birds. In the red grouse from the northern temperate moorlands of the UK and Ireland, both sexes bear combs on their heads that reflect UV light. The strongest reflectance is related to a low parasite load in the owner's body. Heavily parasitized birds produce little UV-reflecting pigment, making the UV

---

[91] Servedio, M.R., and Lande, R. 2006 Population genetic models of male and female mutual mate choice. Evolution 60: 674–685.

reflectance signal a true statement about the bird's state of health. In the pied flycatcher from Europe and Asia, to take another example, the size and quality of the white forehead patch is correlated with low parasite load and, in females, the number of surviving offspring. In both cases, low parasite load means two things. Firstly, it means the individual's immune system is strong, and secondly it means that he or she is more likely to be up to the task of raising the young than an individual carrying a high parasite load. The data show that high-quality mates give rise to more surviving young than low-quality mates, indicating the importance of putative partners reading the signals before they choose their mates.

A couple of examples of where mutual adornments are employed in gregariously living birds may help to establish the principles of mutual ornamentation in monogamy. Birds make good subjects for experimental work because feathers can be cut or augmented, without harm to the individual on whose bodies they grow. One of the best-researched examples of mutual sexual selection has been conducted on the crested auklet, a small colonial sea bird living around the coasts of the Bering Sea. Once a male and female crested auklet have paired at the start of the breeding season, they remain faithful to one another throughout that season, by the end of which time their young are ready for an independent life. At the onset of breeding both sexes develop little forward-falling crests of feathers on the tops of their heads. If the crests of birds are experimentally augmented, by gluing on extra and longer feathers, members of the opposite sex pay them more sexual attention than they do to non-augmented control birds. If they are made smaller, they are paid less attention. These effects are seen in both sexes equally, showing that the crests are for mutual sexual signalling purposes (Figure 8.2).

Crested auklets don't stop at only a mutual visual ornament, for they also have a mutual scent ornament. The birds emit a tangerine-like scent, discernible up to 1 km downwind from their cliff-face colonies. The scent is present only during the breeding season and is produced by glands in the nape of the neck in both sexes. Julie Hagelin of Swarthmore College in Philadelphia observed that males and females engage in what she calls 'ruff-sniff' behaviour, in which

(a)

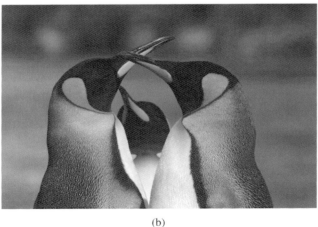

(b)

Figure 8.2

(a) Crested auklet, showing head crest, and (b) King penguins, showing yellow ear' spots and bill flashes, (a) US Fisheries and Wildlife Service, F. Deines, (b) Copyright Ardea Picture Library Ltd.

one bird buries its bill into the feathers at the nape of the other's neck.[92] This behaviour isn't seen outside the breeding season. Short cylindrical neck feathers are modified into wicks that aid in dispersing the scent, much as the hairs in the human armpit disperse scent. Chemical analysis of nape feathers showed the scent to be a mixture of volatile alcohols, acids, and aldehydes, and is the same in both sexes. If crested auklets of either sex are put into a 'T'-shaped maze, down one arm of which flows crested auklet scent and unscented air down the other, the birds move towards the scented arm, and it doesn't matter whether the scent donor is male or female. Stuffed auklets treated with neck scent are approached by both sexes more frequently than unscented models, suggesting that in these interesting little birds there are at least two sensory modalities involved in mutual ornamentation.

Another example of mutual adornment is seen in the striking markings on the side of the head and bill of king penguins from the sub-Antarctic islands. These handsome birds have an extremely low lifetime reproductive rate, producing only two offspring every three or four years, and males and females are incredibly difficult to tell apart. Each chick takes about 14 months from egg to independence, requiring a high commitment from both parents if reproduction is to be successful. Each chick needs to be guarded alternately by one parent or the other against attacks by skuas and giant petrels. Both parents spend time at sea feeding themselves and, later, catching fish for their growing chick, always leaving one parent on guard duty. Both sexes have a patch of bright yellow-orange feathers lying over the 'ear', and an orange-coloured streak along the lower bill, which reflects UV light. Research carried out on Crozet Island in the Southern Ocean has shown the importance of the colourful signals in mutual mate choice.[93] Paul Nolan and his colleagues used black felt-pen marker to reduce the size of the ear patch in some birds, and marine varnish containing ground chalk to cut down the UV reflectance of

---

[92] Hagelin, J.C. 2007 The citrus-like scent of crested auklets: reviewing the evidence for an avian olfactory adornment. Journal of Ornithology 148 (Suppl. 2): S195–S201.

[93] Nolan, P.M., Dobson, F.S., Nicolaus, M., et al. 2010 Mutual mate choice for colourful traits in king penguins. Ethology 116: 635–644.

the bill streak. Both practices resulted in treated birds taking much longer than control birds to find a mate and form a pair-bond. Both sexes find high UV reflectance from the bill to be a mutually attractive feature. Females were considerably choosier than males with respect to the size of the 'ear' patch, a feature that appears to be related to social dominance and a bird's ability to find a good spot in the breeding colony of hundreds or thousands of birds. The pigment causing the yellow-orange colour is a substance that contributes to the competence of the birds' immune system, and thus is an honest expression of the birds' quality. These physical expressions of genetic configuration and overall quality are mutually assessed by both sexes at the start of the longest period of commitment known in birds.

*   *   *

It's clear from the above examples of mutual adornments that the similarities between them and human axillae are striking. All appear at the onset of sexual maturity, placing them firmly in the context of sexual biology. All are equally well developed in both sexes. The most obvious difference between axillary organs and bird adornments is that axillary organs are active the whole year round, whereas in birds the ornaments only develop in the run-up to the breeding season, and are discarded, or moulted-out once it's over. This difference is explained by the fact that humans are sexually receptive and engage in sex year round, while birds have strictly seasonal breeding cycles. Humans have year-round sexual receptivity because human young are dependent for far longer than any bird chick, so the pair-bond must persist for some years.

Studies on the fixation of genes for parental cooperation, together with experimental work on crests and feathers, shows that ornaments used in mutual displays between the sexes play a part in keeping them together. It's postulated that axillary organs evolved in our ancestors, through mutual sexual selection, to assist in strengthening the pair-bond between one male and one female. Their location on the upper body, close to a host of visual sexual releasers, enables their products to be perceived during courtship and sex, contributing olfactorily to Morris' notion of 'personalized sex'. During courtship, the scent from the organs mutually assures both members of a putative pair about each partner's immune system configuration, providing assurance about each individual's genetic compatibility. During our ancestors' evolution, 'incompatible' pairs produced fewer offspring than compatible couples, so strengthening selection for individuals with compatible immune systems to pair. Over time, armpits became another pair-bond adaptation that can now be added to Desmond Morris' list.

Is serving the pair-bond the only role for axillary scent? Perhaps not. In leaf-eating primate species in which a single male dominates a small troop of females and young, fathers usually mate with their daughters when they first come on heat. Humans are a blend of old primate leaf-eating ancestry and new carnivore adaptation, in which a system for the prevention of breeding with one's own close relatives has evolved, a system called 'exogamy'. The genetic dangers of inbreeding are so strong that exogamy is effected by the body scent that signifies the configuration of a person's immune system. As Carole Ober's study on marriage in the Hutterites showed, young men and women marry partners with different immune characteristics to their own, avoiding those with identical configurations, though they do this quite subconsciously. Overnight T-shirt experiments show us such information is encoded in axillary scent. The axillary organ and its scents help to blend our old primate ancestry with our new carnivorous lifestyle, helping to reduce incest and genetic turbulence.

Humans have an unusually high number of genes for immunity, suggesting that immunological competence was very important to the evolution of our ancestors. One hypothesis proposes that early hominoids faced waves of invasions by infectious organisms, with each infection requiring a specific antigen to combat it, a situation that would result in high gene diversity. Whether waves of invasions by pathogens might have been encouraged by our ancestors' communal lifestyles, as we can see today in the spread of epidemics, isn't known.

Genetics tells us a rapid expansion in the number of immunity genes occurred sometime around 17 million years ago — after the separation of hominids from the Old World monkeys, and long before humans appeared. Selection of mates with low pathogen loads would have been particularly important in a species that had a low lifetime reproductive output, and required mate cooperation to successfully rear offspring to independence. The basic elements of the modern immune system arose about 450 million years ago, when the earliest jawed fish swam in the oceans, and when the olfactory system was already well established. In these somewhat simple animals both the MHC and the olfactory receptors would have been under strong evolutionary pressure. Peter Doherty, the Nobel Prize-winning immunologist, posits the question: 'Is the requirement for immune diversity to counter new pathogens so central to the survival of mammalian species that the olfactory link to the MHC class I [genes] has been favoured as a further driver of polymorphism?'[94] The answer would appear to be yes, at least in humans and chimpanzees,[95] though we still don't know why their ancestors should have needed more immune surveillance than other mammals. Whatever the reason, humans of both sexes wear a statement of their immune competence on their arms — or, more precisely, under them — enabling potential mates to subconsciously assess their immune system configuration, and to choose mates with high resistance to pathogens.

*　　*　　*

If axillary organs evolved through sexual selection as mutual adornments to provide information about their carrier's immune system configuration, how can the substantial reduction of axillary smell, characteristic of south-east Asian people, be explained? Does this mean they don't have the subconscious ability to compare immune system configurations of potential partners that people of other races do? The fundamental difficulty here is that it's not known what

---

[94] Doherty, P. 2004 On the nose. Shared themes for the sensory an immune self. Chiron 2004: 9.

[95] Bakewell, M.A., Shi, P., and Zhang, J. 2007 More genes underwent positive selection in chimpanzee evolution than in human evolution. Proceedings of the National Academy of Sciences USA 104(18): 7489–7494.

component of the axillary secretion is responsible for transmitting information about its owner's MHC configuration — indeed, it's likely *not* to be part of the consciously perceived apocrine scent, since apocrine scent varies so widely among races, and there is no evidence of increased immune system incompatibility in populations where perceptible axillary smell is diminished.

The genetic analysis of human origins indicates that the allele to reduce apocrine scent first appeared after the 'out-of-Africa' event, some 100,000 years ago, when human ancestors travelled north from Africa, to colonize the world.[96] The allele is strongly correlated with latitude, suggesting it might be related to an adaptation for survival in a cold climate. Genetic adaptation for survival in a cold climate may have simply swept the axillary allele along as collateral damage, as it were — such things happen not infrequently in evolution. The allele does not completely extinguish apocrine smell, and likely has no effect upon the production of steroids or other constituents that occur at low concentrations. The fact that the axillary scent of Asian people is reduced doesn't mean that information about their MHC configuration has been extinguished; until more information is available about what component of axillary scent carries information about the immune system compatibility the matter can't be settled.

Our evolutionary background, and the importance of mutual adornment within the pair-bond, gives an indication why Huysmans, and other writers rate the armpit so highly. Armpit scent organs evolved in Adam's and Eve's ancestors to enable a subconscious exchange of information about the emitter's immune system configuration and compatibility. Once compatibility or otherwise has been established, the constant presence of the familiar scent helps maintain the bond between the partners, supporting long-term sexual security for the purposes of seeing slow-growing offspring through to independence.

---

[96] Ohashi, J., Naka, I., and Tsuchiya, N. 2011 The impact of natural selection on an ABCC11 SNP determining earwax type. Molecular Biology and Evolution 28 (1): 849–854.

Chapter *9*

Incense and Perfume

So little is known about the development of human culture that it's hardly surprising that nothing at all is known about when, or why, a human being first applied some material to their body *purely for its scent*. It's possible that the scent of some aromatic herb rubbed onto a wound for the herb's healing quality might have been the initial trigger, but this is speculation. This idea isn't as fanciful as it may seem at first glance however, because animals can be observed to self-medicate; a branch of biology called 'zoopharmacognosy' chronicles what is known about it. Many different kinds of animals have been observed to seek out certain herbs that are not at all nutritious but have some medicinal value, such as for purging parasites from the gut, for example. There's anecdotal evidence of baboons eating leaves of the soapberry tree, or Egyptian balsam, to combat the scourge of schistosomiasis (bilharzia or snail sickness), and orang-utans eating leaves of various plants apparently to rid themselves of intestinal worms, but empirical evidence is lacking. Baboons are also thought to eat certain leaves to induce an apparent sense of euphoria, leaves they do not normally eat and that appear to the human observer to make the baboons more alert and active.

Observations of some species of monkeys in the wild, reveals behaviour much more likely to be on a path leading to the evolution of the purposeful application of perfume. Monkeys sometimes chew plant material before rubbing the salivary mess into their fur. Such behaviour has been carefully documented in capuchin monkeys, but is also seen in South American spider and owl monkeys. Mary Baker from the University of California documented this behaviour in capuchins in Costa Rica; the monkeys use the leaves and stems of the white-bearded vine *Clematis dioica* and fragrant pepper *Piper marginatum*, plants not usually included in the monkeys' diet.[97] Local people use these plants to relieve pain, reduce swelling, and relieve fever. She noticed that fur rubbing with chewed leaves was much more prevalent during the wet season when mosquitoes were in murderous abundance than during the dry, and as *Clematis* and *Piper* scents are quite effective mosquito repellents, the behaviour might have evolved for insecticidal purposes. *Clematis* and *Piper* also produce antibacterial and antifungal compounds, so may serve additionally to keep the monkeys' skin free of fungal irritation.

Monkeys have been observed to crush and rub toxic millipedes into their fur to deter insects from biting; aromatic compounds called benzoquinones produced in the millipedes' mouths deter predatory birds from pecking at them, and are also effective insect repellents. Ants too, are used for repelling biting insects by many species of birds and also by a few mammals, including capuchin monkeys and some small rodents. Ants are killed and deposited in the feathers and fur in a behaviour known as 'anting'. The strong smell of formic acid from the crushed ants does the rest. Application of citrus juice, millipede and ant smell, and other plant materials to the fur appears to be employed to make life less pleasant for the legions of creeping parasites, as well as fungi and harmful bacteria that infest everything with a hairy coat.

It's common amongst mammals for individuals to signal social status and sexual condition through their body smell. In seasonally breeding species, and especially those in which males acquire a harem of females, there's much pressure on a successful male to maintain his position, and not be diverted from the job of mass mating by challenges from younger or fitter males. Mass mating places a great strain on a high-ranking male, who may start to lose body condition even after just a few days into the rut. During this time,

---

[97]Baker, M. 1996 Fur rubbing: use of medicinal plants by capuchin monkeys (*Cebus capucinus*). American Journal of Anthropology 38: 263–270.

male goats, reindeer, elk, moose, ibex, and many more species display a behaviour in which they spray their urine onto their chests, necks, and beards. The oldest, biggest, and fittest males have the rankest body odour during rutting time and by spreading the malodorous liquid over their fur, are able to mask their decline in body condition — at least for a few days. But even an extra day or two of retained dominance may equate to a much higher number of females covered, providing adequate selective pressure for the evolution of olfactory deception.

While these are dramatic examples of olfactory deception for gaining a reproductive advantage, less dramatic examples are seen in monkeys. Capuchin monkeys urinate on their hands and feet, and transfer their odour signatures to the branches as they go about their daily business. The trail of scent marks, and the spread of personal information throughout the colony, plays a role in keeping the group socially stable. Captive capuchin monkeys are fascinated by pungent-smelling plants, and particularly by onions, which they chew lightly before rubbing onto their fur. Normally capuchin monkeys are quite sociable towards one another, but quite soon after they've been chewing onions become uncharacteristically aggressive. It seems that the pungent smell of onions masks their urine smell, resulting in a partial breakdown in the troop's social stability. The onion effect soon passes and the monkeys revert to their normal pattern of social interaction. We'll never know how the first use of perfume by early humans came about, or from what functional behaviour it evolved, but the moment one of our ancestors applied scent to his or her body, *for the pleasure of its scent rather than for any other reason*, was the moment our olfactory culture began.

Odour culture in modern human society revolves around incense and perfumes, though there's little real difference between them. For most of recorded history, perfumes to grace the body were made from what were then, as now, incense ingredients. Incense is the name given to plant materials used to perfume the air with scented smoke, and is usually effected through roasting various parts of plants and plant exudates on hot charcoal. Anthropologists speculate that social gatherings of early humans around cooking fires stoked with aromatic woods would have provided the opportunity for the inclusion of the scent of smoke to be incorporated into the worship of a spirit, and might be how incense use originated. The fact is we've no idea what induced an ancestral Robinson Crusoe to wish to consciously change his or her smell, or to enjoy the smell of a fire when stoked with aromatic wood, or even when it occurred. But happen it did, and from that time forward perfume and incense have been integrally associated with human social evolution.

*　　*　　*

The first *recorded* uses of incense come from ancient China and India, where swirling smoke has been used in religious services for the past 7,000 years, and probably for much longer. Perfumes may have also been used from those times, though the first known factory engaged in manufacturing perfumes for widespread use dates from the Bronze Age in Cyprus, some 4,000 years ago. Archaeological work in 2005 uncovered clear evidence of a perfume factory at Pygos, about 60 miles south-west of Nicosia, complete with stills, bottles, mixing bowls, and other tools of trade. Myrrh and frankincense formed the mainstay for both incense and ancient perfumes, together with cinnamon, sweet cane, and the scents of a few kinds of flowers. Extraction of fragrances was brought about through distillation of petals or other parts, or through a process known as 'enfleurage' when the essential oils of the flowers were absorbed into oil or animal fat. Egyptian tomb reliefs showing women and men wearing cones of scented animal fat on their heads, to perfume the space around their faces, are commonplace (Figure 9.1).

The extraction of essential oils of plants with alcohol didn't occur until the 14th century in Europe, with the appearance of Hungary water (rosemary and thyme distilled in brandy), Oil of Neroli (orange and bergamot distillate) and *eau de cologne* (citrus, bergamot, lavender, jasmine, and rosemary distillate) in the late 16th and early 17th centuries. The light, floral perfumes so popular today were unknown to the ancients, though they'd have been a godsend

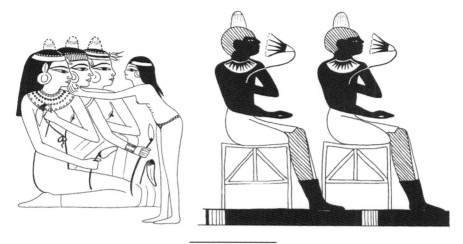

Figure 9.1

Left: A cosmetic slave attends three young Egyptian women. The women are wearing head cones made of animal fat and myrrh. Right: Egyptian men wearing myrrh cones. Both men and women are holding sacred lotus flowers. Reprinted with permission from Stoddart D.M., 1990 The Scented Ape: the Biology and Culture of Human Odour, Cambridge University Press.

to men and women spending long hours under a blazing Egyptian sun with piles of lard on their heads!

The composition of early incense is better recorded than early perfumes, probably on account of its role in religious services. All the world's great religions, except Islam, use incense in their regular worship and have their own formulae for holy incense. The similarity between various incense recipes is greater than their differences. Moses was instructed by God to prepare incense from myrrh, frankincense, onycha,[98] and galbanum (Exodus 30: 34–36). The Torah lists 11 ingredients as essential for incense to be offered in the Jewish

---

[98]There's ongoing confusion about the identity of onycha. It's generally thought to be the resin from a rock rose found widely distributed throughout the Middle East (*Cistus ladanifer*). However, some early reports note it to be the fingernail-like cover (operculum) of marine snails (e.g., *Strombus lentiginosus, Onyx marinus, Unguis odoratus*). As sea creatures were counted among the unclean, it's highly unlikely they would have been used in holy incense.

tabernacle, namely myrrh, frankincense, onycha, galbanum, cinnamon, stacte, cassia, nard, saffron, costus, and cinnamon bark.

Hindu ritual requires use of very complex mixtures of incense that must include stems, roots, flowers, leaves, and fruit. The stems used were myrrh, frankincense, cinnamon, cedar, sandalwood, and aloeswood. The roots were costus, nard, valerian, turmeric, and vetiver. To satisfy the flower requirement, spicy blooms of clove were used and for leaves, patchouli was favoured. The fruit was usually star anise, with its heady smell of dark liquorice. Other ingredients such as cardamom, camphor, and betel (both leaves and nuts) could also be included, depending on local availability. Many ingredients from Taoist, Shinto, and Buddhist incense were introduced to the West as trading routes from China became established.

While ingredients differ across such a vast landmass as that spanned by China with its many climatic regions, the fundamental ingredients of Chinese incense are frankincense, Chinese cinnamon (not quite the same as Middle Eastern cinnamon), cedar, sandalwood, aloeswood, benzoin, nard, cloves, star anise, fossil amber (the Talmud mentions the possible inclusion of 'Jordan amber' in Jewish incense), camphor, fennel, two types of ginseng, laurel bark (possibly another kind of cinnamon), and wild rhubarb. As time progressed and trade between East and West increased, ingredients from the East, particularly sandalwood, were progressively incorporated into incense used in Christian and Jewish worship, and perfume concoctions as shown in Tables 1 and 2.

Moses was instructed to prepare holy anointing oil from myrrh, cinnamon (two kinds), and calamus mixed with olive oil, to be used as perfume for anointing the tabernacle and the Ark of the Covenant.[99] Mostly, and particularly prior to the rise of the Roman Empire, incense and perfumed anointing oils were only used in religious services and were not available to ordinary mortals, however much they were wanted for their evocative and sensual qualities. They were holy, to be used only in religious ritual for accompanying prayers to the gods and there were strict spiritual penalties for secular use.

---

[99] Book of Exodus Chapter 30: v 23–35.

Table 1. The main ingredients of incense used by the main religions in their holy rituals.

Name	Alternative names	Type of substance	Species	Notes	Christian incense	Hebrew incense	Chinese incense	Hindu incense
Myrrh	Balsam; Balm; Bdellium; Stacte (oil of myrrh); Murru; Nataf	Resin	*Commiphora myrrha*	Regarded as the king of scents, admired above all others	✓	✓		✓
Frankincense	Olibanum; Luban; Lebonah; Tus	Resin	*Boswellia sacra*	Universal incense	✓	✓	✓	✓
Onycha	Labdanum; Ladanum	Resin	*Cistus ladanifer; C. creticus*	Balsam-like scent. The essence of Chypre	✓	✓		
Galbanum	Asafoetida; Devil's Dung	Resin	*Ferula galbanifula*	Pungent scent	✓	✓		
Cinnamon	True cinnamon	Bark	*Cinnamomum verum*	Balsam-like smell; undertones of bergamot and orange		✓	✓	✓
Cassia	Chinese cassia; Chinese cinnamon	Bark	*Cinnamomum cassia*	Balsamic smell		✓	✓	✓

(Continued)

Table 1. (Continued)

Name	Alternative names	Type of substance	Species	Notes	Christian incense	Hebrew incense	Chinese incense	Hindu incense
Calamus	Sweet cane; Sweet flag	Root	Acorus calamus	Tangerine-like smell		✓		
Sandalwood	Algum; almung wood; sandal	Wood	Santalum alba	Faintly urinous smell			✓	✓
Costus	Koshet	Root	Saussurea costus	Smell of Chinese joss sticks		✓		✓
Nard	Spikenard; Sweet rush	Whole plant	Nardostachys jatamansi	Faintly musk-like smell; hint of animals		✓		✓
Saffron	Crocus	Pollen	Crocus sativus	Sweet smell		✓		
Benzoin	Gum Benjamin; Gum benzoin	Resin	Styrax benzoin	Balsamic smell			✓	
Cedar	Incense cedar	Wood	Juniperus virginiana	Faintly musk-like; hint of civet smell			✓	✓
Star anise		Fruit	Illicium verum	Aniseed/liquorice-like			✓	✓

Name	Other names	Part used	Scientific name	Smell		
Aloeswood	Agarwood, Eaglewood; Lignum aloes; Oud; Agar	Fungus infested heartwood of several species of tree	*Aquilaria malaccensis*	Complex and very pleasing, earthy smell	✓	✓
Camphor	Camphire; Henna	Bark	*Cinnamomum camphora*	Balsamic smell	✓	✓
Clove		Flower and fruit	*Syzygium aromaticum*	Earthy, spicy, slightly sweet	✓	✓
Patchouli	Patchouly	Leaf	*Pogostemon cablin*	Very strong, earthy, woody smell 'like lavender's unwashed cousin'		✓
Ginseng	Lovage; Mountain ginseng; Siberian ginseng	Root tuber	*Ligusticum wallichii*; *Eleurothercoccus senticosus*	Resembling liquorice, sarsaparilla, anise and cloves with sweet undertones	✓	

(Continued)

Table 1. *(Continued)*

Name	Alternative names	Type of substance	Species	Notes	Christian incense	Hebrew incense	Chinese incense	Hindu incense
Fennel		Leaf and stem	*Foeniculum vulgare*	Anise-like			✓	
Indian rhubarb		Leaf stem	*Rheum officinale*	Bitter astringent smell			✓	
Borneol		Resin	*Dryobalanops aromatica*	Balsamic smell				✓

Human beings are able to live their lives on both sacred and earthly planes, and it's easy to see how they would have wanted the intoxicating scent of incense experienced during worship to decorate their earthly lives, just because it has a nice smell. During the Greek and Roman Empires, a perfume culture quickly arose and factories scoured the land for the most powerful ingredients. Pliny the Elder tells us that Julius Caesar issued an edict in Rome that perfumes weren't to be sold to ordinary folk, but his words fell on deaf ears and demand for 'exotics', as perfumes from the East were known, soared. Several distinct recipes were compounded, doubtless influenced by the local availability of ingredients.

Myrrh and cinnamon, together with frankincense, were the backbone of early perfumes. Other scents, including the fragrance of the sacred lotus flower, would have provided lighter notes. The perfume of roses was very popular in ancient Persia and is inextricably interwoven with the Islam and Hindu religions. It's said that when the Prophet Mohammad died in the year 632, a drop of sweat fell from his brow as he ascended to heaven, and from it sprang up a rose bush. The scent of the rose became the scent of the Prophet. Rosewater, or rose attar, is used extensively to this day in Islamic life. There's evidence from ancient Indian manuscripts that rosewater was first used thousands of years before the Prophet's ascension. Rose scent in the perfumes of Rome and Greece in antiquity almost certainly owes its existence to its use in ancient Persia.

The recipes for a few Egyptian, Greek, and Roman perfumes have come down to us through the writings of Theophrastus and Pliny. '*Megaleion*', named for the great Greek physician Megallus, '*Egyptian*', '*Susinum*', and '*Mendesian*' were the hits of the time. Sometimes '*Kyphi*' was included as a perfume, though mostly its use was restricted to temple ceremonies. '*Susinum*' was considered the most appropriate for men's use as it was a powdery perfume that could be sprinkled onto the beard, or rubbed onto the bed covers to impart a more distant fragrance. Jars to hold perfume were frequently beautifully fashioned and decorated to show what was within — just as are perfume *flaçons* today.

Table 2.   The main ingredients of some ancient perfumes.

Ingredient	Megaleion	Egyptian	Susinum	Mendesian	Kyphi
Myrrh	✓	✓	✓	✓	✓
Cinnamon	✓	✓	✓	✓	✓
Frankincense	✓	✓			✓
Cardamom			✓	✓	
Water lily (lotus)		✓	✓		
Rose		✓	✓		
Cedar		✓			✓
Labdanum		✓			✓
Saffron			✓		
Bitter almonds				✓	
Juniper					✓
Benzoin					✓
Calamus					✓

Occasionally the oily distillate prepared from the roots of a grass known as vetiver, or 'khus' in its native India, *Chrysopogon ziza-nioides,* was added as a perfume fixative, to make the scent last longer. This was especially the case if the perfume was to be stored for any length of time.

*   *   *

The question arising from all this is what is so special about these scents that one species of ape, recently evolved from the grassy plains of Africa, should have developed an entire odour culture around them? What is so special about this suite of plant products and why is the range of ingredients so limited? The Christian prophet Jeremiah asked much the same question: 'To what purpose cometh there to me incense from Sheba and the sweet cane from a far country?',[100] probing the practical, the religious, and the profane. Smoke has

---

[100] Jeremiah Chapter 6: v 20.

served to cleanse the body, ridding it of parasites adhering to the skin and hair from the time fire was first used for cooking. To this day, smouldering mosquito coils, slowly burning to release a fragrant smoke, are quite effective in driving away insects looking for a feed. Fleas, lice, ticks, and mites don't like smoke either and can be purged by the body's immersion in smoke. Ancient Egyptians knew of the insecticidal properties of frankincense smoke, fumigating their wheat silos with it to rid them of grain beetles and moths. From this use of smoke in cleansing the body and larders, it's likely that it took on the role of cleansing the mind, and thus it *segued* into religious service.

Although incense has been used in religious ceremonies from the start of recorded history, there have been quite long periods of time when its use declined, only for it to then gradually creep back up again. In early Christianity, incense was banned from use as its association with Hebrew worship, and Judas' betrayal of Christ, was too strong. It took 400 years before it reappeared in the Christian church, when Constantine the Great ushered in the Peace of the Church, and today in some Christian sects its use rivals the profligacy of Egypt's New Kingdom, 3,500 years ago. But the smell of incense is associated with something far more basic and biological, stirring the animal from which the religious ape evolved.

*   *   *

The ancients knew perfumes had erotic characteristics that could arouse the perceiver. Cicero spoke against the 'spreading of ungents', and odorous and erotic preparations on the person, while Ovid advised young Romans to avoid perfumes and cosmetics, as they were the province of prostitutes and not fit for proper citizens. The long-held association of perfumes with prostitutes is reflected in the French word for prostitute *'putain'* and the Spanish *'puta'* — both words derived from the Latin verb *'putere'*, meaning 'to rot', from which the English 'putrid' comes, with all its connotations of foul smell and unwholesomeness.

The Greek herbalist Dioscorides was the first to record the mood-changing properties of perfumes, noting that costus and crocus 'stir up venery', though such properties have been known and employed by shamans and mystics in many cultures from ancient times. Many plants have psychoactive odours that can change a perceiver's mood. Mystics and shamans induce entrancement and hallucinations after deeply breathing smoke from the burning of various plants and herbs. Heightened sensations follow. Cannabis smoke carries agents of psychological and physiological activity from the lungs into the bloodstream, and perhaps its smell also contributes something to its narcotic effect. Incense smoke has long been thought to exert a psychoactive effect and quite recently that claim has been substantiated.

To understand how incense may alter one's state of mind, Arieh Moussaieff from the Hebrew University in Jerusalem recently reported isolating a compound in frankincense that he named 'incensole acetate'.[101] This compound switches on a receptor that relaxes blood vessels in the skin, creating a feeling of gentle warmth as more blood courses closer to the skin's surface. The receptor, which goes by the name of Transient Receptor Potential Vanilloid 3, or TRPV3, is expressed mainly in the skin but it also expresses in the brain. Moussaieff's team conducted a series of biochemical and genetic studies on mice in which this gene was switched off, using as controls mice with it left to function normally. They showed that normal mice displayed far less anxiety in standard maze-running tests when injected with incensole acetate than when they were injected with normal saline as a control substance. Mice lacking the gene showed no diminution of their anxiety when injected with incensole acetate, demonstrating that the compound is able to alter a normal mouse's state of mind only if it can switch on TRPV3 in the brain. The parts of the brain activated by incensole acetate are the amygdala and septum, together with the motor cortex. Moussaieff's team concluded that the active ingredient in frankincense stimulates the emotions by reducing anxiety and depression, as well as by physiologically inducing a feeling of bodily warmth.

---

[101] Moussaieff, A., Rimmerman, N., Bregman, T., *et al.* 2008 Incensole acetate, and incense component, elicits psychoactivity by activating TRPV3 channels in the brain. The FASEB Journal 22: 3024–3034.

Incensole acetate hasn't been isolated from any species other than frankincense; more than 50 traditional herbal plants from the orient have now been examined and no trace of the compound has been found in any of them. Neither has any other compound able to switch on TRPV3 been found. This is why frankincense is so special and perhaps explains why it is such a universal ingredient in incense and perfumes. The similarity with the manner in which the human mood-changing pheromone, androstadienone, alters one's emotional state is striking. Perhaps we now know the answer to Prophet Jeremiah's conundrum about incense; it doesn't only have a nice smell — it does nice things to your mind.

Jeremiah may have been the first to question the growing use of incense in religious ceremonies, but he wasn't the only sceptic. In Tunis, the great scholar Arnobius of Sicca, a staunch defender of early Christianity prior to Constantine mending the rift between Judaism and Christianity, remarked severely[102]:

> 'What is this sign of respect which comes from the smell of gum of a tree burning in a fire? Does this, do you suppose, give honour to the heavenly magnates? Or if their displeasure has been aroused at some time, is it really soothed and dissipated by incense smoke? But if it is smoke the gods want, why do you not offer them any kind of smoke? Or must it only be incense? If you answer that incense has a nice smell while other substances have not, tell me if the gods have nostrils, and can they smell with them? But if the gods are incorporeal, odours and perfumes can have no effect at all upon them, since corporeal substances cannot affect incorporeal beings.'

We now have the answer. Incense is burned not to take prayers up to the gods, or to represent the odour of sanctity, but to induce warmth and a relaxed state of mind in the congregation by reducing levels of anxiety. In such a state, the mind is opened to the emotional persuasion and suggestion upon which faith depends.

*　　*　　*

---

[102] Arnobius of Sicca *Adversus Nationes* Internet Archive. Available at: archive.org/stream/thesevenbooksofa00arnouoft/thesevenbooksofa00arnouoft_djvu.txt.

Some resin alcohols found in incense have structures that resemble steroid hormones in animals. As we saw in Chapter 3, similarity in chemical structure is no predictor of smell similarity so we must tread cautiously. Ancient writings, though, are full of references to the seductive power of incense ingredients — the writer of the passage describing the end of adultery declares[103]:

> 'Therefore I have come forth to meet you, diligently to seek they face, and I have found thee. I have decked my bed with coverings of tapestry, with carved works, with fine linen of Egypt. I have perfumed my bed with myrrh, aloes, and cinnamon. Come, let us take our fill of love until the morning: let us solace ourselves with love.'

When the young virgin Esther was being prepared for marriage to King Xerxes she had to undergo lengthy purification[104]

> '...to wit, six months with oil of myrrh and ...other things'

and we can recall the elaborate preparations undertaken by Queen Cleopatra before she met Marc Antony, with the steely objective of seduction in mind.

The German perfumer Paul Jellinek questioned whether incense ingredients impart erotic measures to light floral perfumes.[105] He spent a lifetime working in the perfume industry and was surrounded with professional 'noses', as those gifted people skilled in the art of blending fragrances into perfumes are known. He asked his 'noses' to describe the scents of frankincense, myrrh, styrax (Balm of Gilead), cinnamon, labdanum, sandalwood, benzoin, costus, and sweet cane, and found that with the exception of sweet cane, all reported they introduced an 'erotic' dimension to floral fragrances. While 'noses' are highly trained at articulating what they smell, they are but normal human beings living in the normal world, and subject

---

[103] Proverbs Chapter 7: v 15–18.
[104] Book of Esther Chapter 2: v 12.
[105] Jellinek, P., and Jellinek, J.S. 1994 *Die psychologischen Grundlagen der Parfümerie*. Hütig Buch Verlag, Heidelberg.

to everything to which the rest of us experience. While it's entirely possible that their observations were based on their experiences of life, rather than on some fundamental quality of the incense ingredients, their articulations of scent quality indicate that incense ingredients are reminiscent of animal smells. Some of the panel went a little further and equated myrrh with the smell of fair-haired people, frankincense with dark and red-headed subjects, styrax with the skin of dark-haired people, and costus with human armpit scent.

In another study, the Dutch psychologist Jan Kloek asked 100 students to sniff a range of steroid chemicals and report what they smelled of.[106] Over a quarter reported a woody or sandalwood smell, though to be fair it must be said that the majority of subjects could only say they could smell something, but were unable to give it a name. Next to the recognition of sandalwood was the recognition of urine. The active ingredient in the smell of sandalwood is androstenol, a steroid found, as we know, in human urine and armpit scent.

These studies indicate that incense ingredients have smells that the brain recognizes as resembling smells of the human body. Perfumes in ancient times were compounded mainly from incense ingredients and thus emitted strong human–animal smells. Were they worn to mask the lack of running water, soap, and flush toilets, or to enhance the wearer's natural body smell? What we know of Cleopatra's arrival at Tarsus tells us that perfumes were worn then as they're worn now, to advertise one's presence and to make the wearer identifiable by that advertisement. Times have changed only inasmuch as modern perfumes contain more light floral and fruity scents than was previously the case. Our first impressions on meeting a perfumed stranger today are essentially the same as they would have been 3,000 years ago.

It would be nice to know what Cleopatra or Marc Antony would make of modern perfumes. They would probably find them

---

[106] Kloek, J. 1961 The smell of some steroid sex hormones and their metabolites: Reflections and experiments concerning the significance of smell for the mutual relation of the sexes. Psychiatria, Neurologia, Neurochirurgia 64: 309–344.

unsatisfyingly light and trivial in comparison to the scents to which they were used to, which were, in effect, concoctions of liquid incense. Modern perfumes are composed of three main messages or 'notes', to use the musical metaphor employed to describe perfumes (note how the word-rich English language is devastatingly deficient when it comes to smells). The so-called base-notes recall mammalian sex attractants, usually with urinary or faecal reminiscences. The mid-notes are scents of resins and sex steroids, and the top-notes carry the sexual attractants produced by flowers for luring insects to their nectar. Like a good newspaper article, the top notes grab the headlines and turn heads, while the real message is carried in the small print beneath, for the theme of the perfume is encoded in the mid- and base-notes. Today, perfumers equate one's hair and skin colour to certain types of perfumes — a reddish-blonde will, according to Paul Jellinek, harmonize with a stimulating perfume containing cinnamon, mimosa, or hawthorn. A dark-haired or brunette person is best suited to a narcotic perfume containing magnolia, rose, or violet. In Cleopatra's Egypt, variation in hair colour was rare. Dark hair and dark skin lend themselves to heavy perfumes, with strong animal-like ingredients based on myrrh, aloeswood (oud), and patchouli.

*   *   *

We've seen repeatedly that sex is never far from perfume. During the reign of Louis XVI, women courtiers underwent a ritual practice of perfuming themselves that was almost religious, to disguise the fact that they were perfuming their bodies to allure and attract the opposite sex. At the same time, the masses were being discouraged from using musk to perfume their bodies, because strong animal smells confronted the nostrils of genteel citizens. Musk did nothing to lift humankind from its animal origins. But the world has never been short on hypocrisy. In the boudoirs of the rich, hundreds of bottles and vials of perfumes, tinctures and unguents were arranged on an altar, to which a white-robed woman would approach. With due solemnity, attendants would apply a wide range of smells to various parts of the woman's body, reminiscent of how ancient Greeks applied different perfumes to elbows, chest, neck, and hair. The ritual nature of the practice was all that separated the genteel from the masses.

Richard Stamelman, who has thoroughly researched the use of perfumes from the mid-18th century, notes that the eros of scent is simultaneously carnal and spiritual, enabling a body to be both things in a beholder's eye.[107] Perfumes must, he writes, 'give voice to the body', noting that the French Council on Perfume in the 1980s declared '*Sans parfum, la peau est muette*' ('without perfume, the skin is silent'). The message of the voice, however, must be ambiguous. 'Through the eros of scent a woman becomes a hybrid, a hyphenated being: visible–invisible, proximate–distant, corporeal–spiritual, ephemeral–enduring, earthly–celestial, base–sublime, primitive–civilised, expressive–mute, innocent–seductive.' The marketing of fragrances has played on these ambiguities, tapping into, as Stamelman says, 'the reservoir of erotic associations that scent possesses'. Prior to the Second World War perfumes had openly suggestive names, such as *Eros*, *Sèduction*, *My Sin*, and *Tabu*. After the war, perfumes were given more subtle names, suggesting not just sex but a sense of expectation, or self being, such as *Boudoir*, *Intimate*, *Volupté*, and *Basic Instinct*.

Quite recently, and not out of step with the general sexualization of the modern world, some perfumes are being marketed with names designed to extinguish all subtlety of ambiguity. According to '*Elle*' magazine, perfumers are putting fragrances onto the market with names designed to shock, by linking the perfumes directly with human bodies and their fluids, and what humans get up to. Names such as *Putain des Palaces* (*Eau de Hotel Slut*), *Magnificent Secretions*, smelling of sweat, saliva, blood, metal, and semen, *Strip* designed to 'release the sexual attraction within [your body]', and *Narcotic Venus* being 'the result of a quest for the overwhelming addictive intensity of female sexual power'.[108] The canny staff writer

---

[107] Stamelman, R. 2006 *Perfume: Joy, Obsession, Scandal, Sin: A Cultural History of Fragrance from 1750 to the Present*. Rizzoli International Publications, Inc., New York.

[108] Bullock, M. 2007 Dirty pretty things. Elle Magazine. December 2007: 264–269.

for 'Elle', on assignment around the perfume stores, notes that the names on the *flaçons*, and the PR hype, work for 'just for as long as it takes for you to answer the question "credit, or debit?"' How long these hard, blunt, and confronting names will remain on the market remains to be seen, though we should note that the temptation to identify with what others *might* be doing is no different now than it was in the 1950s, when Chanel's fortune was made by Marilyn Monroe declaring she wore nothing in bed except two drops of *Chanel No. 5*!

*   *   *

Exactly when animal products came to be used in perfumes isn't known. It's reported that ancient Indian perfumes included various glandular animal products in perfumes made from leaves, flowers, fruits, barks, woods, and resins. The perfume-savvy ancient Greeks and Romans used only plant products, with the unlikely and contested exception of the fingernail-like closing discs found in a few marine snails. The single most important animal product to find its way into perfumes is musk, produced by rutting male musk deer from eastern Asia and the Himalayas. Musk has been used in Ayurvedic medicine since 5,000 years ago for cardiac, neurological, and mental conditions. The first mention of it as a perfume ingredient is found in the writings of 6th century Greek explorers to India, who noted its purported aphrodisiacal properties. Musk was used extensively in the Abbasid Empire of Arabia, with the caliphs of Baghdad using it lavishly. By the 9th century it found its way to the West as Arab traders reached Europe, and soon no perfume would be made without it.

Animal products are used for four characteristics; they project the perfume away from the wearer filling all the available space with the scent; they import an animal dimension to a perfume; they enhance the penetrating power of the perfume, meaning that the perceiver's sensitivity to it is lowered; and they 'fix' the floral and other ingredients, meaning that they inhibit the evaporation rate of the fragrance. Four animal substances are essential ingredients in the best of today's perfumes, though in the 21st century synthetic molecules have replaced them. They are ambergris, often referred to simply as 'amber', musk, civet, and castoreum.

Amber is produced in the intestines of sperm whales, and possibly some other whales, and expelled with the faeces, though there are popular and unsubstantiated reports of it being vomited up, and expelled from the mouth. It's a grey, greasy gelatinous substance, found floating in the sea, or washed up on the shore. When fresh, it has a disagreeable sewer-like smell, but as it ages it acquires a warmer earthy or woody aroma. It's said to impart a 'velvety' texture to perfumes, enhancing the perfume's theme. The active ingredient is ambroxan, now available synthetically.

Amber was known by the ancient Chinese and ancient Egyptians, who ate it as an aphrodisiac and general cure-all. It's reported that King Charles II's favourite dish was eggs and ambergris! Natural ambergris fetches a high price in today's market. A New Zealand man recently stumbled across an 88 pound (40 kg) piece of ambergris while walking his dog along a beach, reportedly finishing up $400,000 the richer for his lucky find!

In past times, the highest mountain range in the world was home to the musk deer, spanning Afghanistan through the Himalayan nations to India, and east to include much of China. Unfortunately, the little deer have been hunted almost to extinction for the musk pods present in adult males. The word 'musk' comes from the Sanskrit 'muṣkā' meaning testicle, indicating that its relationship with sex was known from ancient times. Musk was so prized in ancient Persia that it was mixed into the mortar used in the mosques at Kara Amed (now Diyabekir, in Turkey) and Tabriz in Iran, so as the sun warmed the buildings' walls the air was filled with the fragrance of paradise. It has a sweetish animal scent, said to be not unlike the smell of a baby's skin, and can be used on its own as a perfume, or to impart lavish warmth to a complex perfume. Concentrated musk has an overpoweringly animal smell, with references to sweat, urine, and faeces but dilution brings a sweetness that imparts an almost mystical quality to perfumes. It never loses its connection with the raw smell of animals.

Today, musk deer are farmed in India and China, though they don't easily breed in captivity. Musk deer are solitary animals, browsing and grazing in dense forest clinging to the foothills of high mountains. During the breeding season, the male's musk gland develops and he smells strongly of musk. Like most solitary animals, mating occurs only when the male and female are attracted to one another, and can overcome their innate fear of contact. Musk does both these things; attracting the female and calming her for long enough to let the male mount her. The pod lies on the male's belly, immediately in front of the penis, allowing some of its precious contents to be squeezed onto the female's genital region to advertise she has been covered, and that no other male needs waste his time (Figures 9.2 and 9.3).

Musk is incredibly valuable. Each pod contains about 25 grams, for which a hunter can get US$300 on the black market. With musk fetching $40,000 to $50,000 per kilogram, there is a high incentive for poachers to kill every deer they encounter. As male and female deer are quite similar in appearance many females are needlessly killed, hastening the species' demise.

Throughout history, unscrupulous dealers have attempted to increase their profits by adulterating musk with other substances, such as congealed blood and dark ochre. Writing in 1742, the English naturalist Charles Owen, quoting Sir John Chardin's travels to Persia some 20 years earlier, tells us how to detect the adulterated product: 'Tis easily counterfeited, and the best way to try it, is by drawing a Thread, dipt in the Juice of Garlick, through the bag [of musk] with a Needle; and if the Garlick loses its Scent, the Musk is good.'[109] Human vanity is pushing many animals to extinction — think of elephants and rhinoceroses pursued for their tusks and horns, and birds of paradise for their beautiful feathers — but the musk deer is the only one to be teetering on the brink because of its smell.

---

[109] Owen, C. 1742 *An Essay Towards a Natural History of Serpents*. London. Published privately for the Author.

Figure 9.2

a) Musk deer; b) civet; c) beaver a), b) and c) Copyright Ardea Picture Library Ltd.

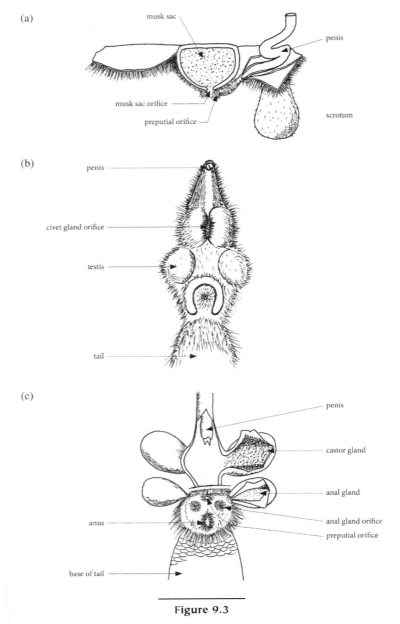

Figure 9.3

a) the musk deer's pod; b) a male civet's scent glands; c) male beaver's castor glands. Redrawn with modifications from Greenish H. G., 1920 A Text Book of Materia Medica. J and A. Churchill.

Nobody in their right mind would wish to smell of the rank, ammoniacal African civet. Related to the mongoose, though frequently — and wrongly — called a cat, the civet's eponymous whitish-yellow paste that's the source of its smell is made in a pair of glands situated in the mid-line of the body, between the penis and the anus (Figures 9.2 and 9.3). Civet serves to mark ownership of a territory and is applied to vertical objects, such as trees, bushes, and rocks. During marking, the orifice of the gland is opened as the civet moves backwards onto the surface to be marked, applying the opening of the gland to the structure below it. As it dries, the deposit turns to yellow and then brown. It remains perceptible to the human nose for several months. When greatly diluted, civet smells alluringly floral-musky, and has long been used as a perfume on its own for its reported aphrodisiacal properties, as well as for perfuming the delicate leather gloves so loved by 19th century European ladies. Petrus Castellus, a writer in the 17th century, notes in his treatise *De hyaena odorifera* that:

> '[civet] will cause so much desire for coitus that she will almost continually wish to make love with her husband. And in particular, if a man wishes to go with a woman, if he shall place on his penis of this same civet and unexpectedly use it, he will arouse in her the greatest pleasure.'[110]

When King Lear declared: 'Give me an ounce of civet, Good Apothecary, to sweeten my imagination', [111] he was requesting undiluted civet, used in Tudor England for relieving depression and providing stimulation. Perhaps this is why the favourite perfume of Catherine Parr, Henry VIII's 6th and last wife, who surely must have had concerns for her own longevity, was civet and juniper.

A male civet produces about 6.4 g of secretion every 5 days, or 32 g per individual per month. Civet can be sold for up to $450 per kg and 1 kg can be used in up to 3,000 l of perfume. While there's a

---

[110] Stoddart, D.M. 1999 *The Scented Ape; The Biology and Culture of Human Odour*. Cambridge University Press, Cambridge.
[111] Act IV scene VI, line 133.

small amount of civet farming in Ethiopia, the introduction of artificial civet is ending the somewhat stressful practice of scraping the civet glands with a spoon, to scoop out the viscous liquid. As an aside, coffee connoisseurs claim that 'Kopi Luwak', or 'civet coffee', is the world's best. A close relative of the African civet, the Indonesian palm civet has a liking for the fruits of the coffee plant, digesting only the soft outer coating and passing the seeds in the faeces, which we know as beans. After drying and roasting, the beans can be sold for $400 a pound, making for an expensive, and well-travelled, cup of coffee!

The last animal substance widely used in perfumes is the product of the paired scent glands of the beaver, known as castoreum or more simply as 'castor'. Castor consists of a yellowish paste-like secretion that's mixed with urine and used for territorial boundary marking. Adult males undertake more territory marking than females and are much more attracted to scent marks, to which they frequently respond by over-marking, but both sexes have castor sacs. Castor was taken internally as an antispasmodic and general sedative until the 17th century, and is not reported to have any aphrodisiacal effects. In the early days of book printing, leather bindings were treated with castor as part of the tanning process. Its scent is known to users of all great libraries around the world, and can be found in leather-themed perfumes and soaps. Fresh castor has a sharp scent, rather like very strong ink, or the resin from birch trees, but when diluted exudes a warm, luscious animal scent with fruity overtones. The secretion is produced in paired castor glands that open into the urethra in both sexes of beavers, and are quite distinct from paired anal glands that sit either side of the anus (Figures 9.2 and 9.3). Castor is still sold on the open market with prices ranging from $30 to $70 per pound at today's prices, depending on quality. Each beaver carries castor worth about $7.50.

Thankfully, from an animal conservation standpoint, these natural products are now artificially synthesized for use in the fragrance industry, though a small amount of the natural products are still used in the most expensive perfumes. Consumer pressure is ensuring that companies find alternatives to using wild animals and this trend is set to grow. Animal scents are added to perfumes to impart musky animal notes without making them smell of urine or the faecal smell of zoos, for humankind doesn't find strong animal smells attractive. There are a number of plants that smell of animals and are likewise employed for their scents. Jasmine's heady scent is due to indole, a compound produced during the decomposition of proteins found in the faeces of carnivores. At very low concentrations, indole has a fresh smell, like citrus or orange blossom. Truffles and morels too, exude the scent of armpits, which at low concentrations are more warm than hostile. Truffle growers use specially trained dogs, or less commonly pigs, to sniff out the underground fungi, but doubtless could do so by themselves.

Our paranoia about body odour comes from the fact that we dare not actually smell of the animal we are, because at high concentration animal smells are overpowering and unpleasant, but neither do we want to smell only of rose and lavender water. Perfumes add complexity to our natural odour by creating something akin to a chord played on an infinite keyboard. Who we are is the complete chord, and not any of its components — many of which we share with other people anyway, much as composers share the use of hundreds of musical tones in their compositions, yet each piece of music is quite individual. Our own signatures are ours and ours alone. Part of our smell is dictated by our genes and the control they exert over the activities of our bodies' scent glands, and part comes from cosmetics containing something from (or artificially synthesized to smell like, as is more likely in current times) real animal bodies. We display our animal origins in the musks and sweaty smells we bring in from outside, keeping our natural animal smell scrubbed away, out of sight and under control.

*   *   *

One last thought about perfumes. You might have on your dressing table, or bathroom shelf, bottles of perfumes bought for you as gifts by other people, and who can resist the beautifully crafted *flaçons*, elegant stoppers, and expensive wrappings. The basis for the giver's choice of fragrance might have been because it reminded them of a former lover, or the perfume promoted by a celebrity they admire and want you to be like. Nice as such gifts are, there's an important reason why you alone should choose perfumes for your personal use, as the advertisement you wish to present must reflect *your* immune system's configuration.

You recall earlier we saw that we are attracted to potential mates with an MHC configuration that's not identical to our own, and that MHC characteristics are subconsciously reflected in armpit odour. Manfred Milinski and Claus Wedekind of the University of Bern presented a range of perfume ingredients to panels of male and female university students.[112] The experimenters knew the characteristics of each person's MHC genotype from a blood test, though the students did not know this for themselves. The ingredients they were asked to evaluate ('Would you like to smell like that yourself?' and 'Would you like your partner to smell like that?') were 36 scents commonly used in perfumery including myrrh, frankincense, cinnamon, labdanum, sandalwood, styrax, benzoin, musk, bergamot, and cardamom, as well as some floral scents including jasmine, orange, violet, rose, lilac, and geranium. Scents were presented on sniffer sticks attached to the lids of small vials to avoid finger contamination, and testers had to rate the ingredients according to the evaluation questions.

The experiment was repeated two years later and the results were quite extraordinary. There was a significant correlation between a responder's MHC configuration and the rating of ingredients he or she would like to encounter in a perfume for self-use over the two-year period, *but no correlation at all for a perfume he or she would like a partner to use.* In other words, the configuration of a person's MHC drives their choice of what they themselves would wish to smell like, but has no influence on their choice of a perfume for someone else. Your partner's choice of a perfume for you is driven by *their* MHC configuration and not yours; if your partner's MHC composition is not identical to yours, as will most likely be the case, you would have chosen something else. This may help explain the

---

[112] Milinski, M., and Wedekind, C. 2001 Evidence for MHC-correlated perfume preferences in humans. Behavioral Ecology 12 (2): 140–149.

prevalence of unsuitable perfumes encountered in supermarket queues and restaurants, and that collection of redundant perfume bottles. Clear them out and choose for yourself!

In an experiment designed to test whether your choice of perfume for personal use increases the attractiveness of your body's natural smell, or simply masks it, Pavlina Lenochová and colleagues asked a group of male volunteers to collect underarm odour using the familiar absorption pad technique, with the caveats on diet, perfumed soap, and sleeping arrangements that we've met before.[113] All odour donors were regular users of a fragrant underarm deodorant, with half of them reporting daily use. The researchers assigned a standard perfume to all donors with the instruction to spray a quick burst of it to one armpit and to use their own preferred perfumes chosen by themselves and not gifted by someone else — on the other. A panel of female students assessed each odour pad on scales ranging from 'unpleasant' to 'pleasant', and 'unattractive' to 'attractive'. The outcome was that axillary odour plus the man's *own selected perfume* was rated significantly more attractive and pleasant than axillary odour treated with the standard fragrance chosen by the researchers. It seems that in making their choice of underarm fragrance, the men had subconsciously selected fragrances that interacted well with their natural body smell, even enhancing its attractiveness, and was not employed to mask their natural body scent. Lenochová's study, together with Milinski and Wedekind's indicate that the perfumes we choose are those that enhance our own odour envelope and scent aura, as much as a well-cut dress or suit enhances our visual appearance. They help us to make a good first impression on our fellows.

Your own smell, then, is as much a part of you as any of your body's physical attributes. While you choose what aroma you would like to define you, it turns out that your perfume isn't entirely a

---

[113] Lenochová, P., Vohnoutová, P., Roberts, S.C., *et al.* 2012 Psychology of fragrance use: perception of individual odor and perfume blends reveals a mechanism for idiosyncratic effects on fragrance choice. PLoS ONE 7 (3): e33810 doi: 10.1371/journal.pone.0033810.

matter of free choice, for subconsciously you select a perfume that enhances, rather than masks, your natural scent. How the configuration of your immune system influences you to make a particular choice isn't known, just as it isn't known how your immune system configuration might influence your choice of life-partner.

Incense and early perfumes were originally compounded from whatever fragrant materials could be found locally. In the Middle East, frankincense, myrrh, cassia, and cinnamon were the essential ingredients; in the Far East, and before the establishment of reliable trade routes, sandalwood, aloeswood, cloves, and star anise were staples. By the Middle Ages, Eastern products had infiltrated Western perfumes, along with glandular secretions from the bodies of animals, that had for so long been at the heart of smoky, animal-like Eastern fragrances. It can be confidently predicted that the perfume world will continuously change, as new synthetic fragrances are developed. The public's appetite for novel smells is as powerful as its love of ancient fragrances, and well weathered by the test of time.

Chapter 10

Smell and High Culture

No account of the making of humankind can be complete without a consideration of how artists and writers deal with the sense of smell, and how they communicate hidden emotional qualities in their work to others. Smells themselves have not traditionally been the subject of cultural exhibition, though there is a contemporary trend for smell-art installations which appear to be gaining a wider acceptance in modern art circles.[114] Artists appeal to the emotions through their creative works by drawing out experiential feelings of happiness, sadness, anger, or arousal as the subject may dictate. Pictures and stories of horrifying or arousing scenes, that would send us running for cover if encountered in real life, release only emotional responses, allowing our experiences and emotions to do the rest.

My purpose in setting smell within the context of high culture is simple; I wish to examine olfactory imagery in artistic expression from the biological perspective of what we know about how smells affect the brain. My objective isn't to analyse olfactory imagery in the poetry of John Donne, or Baudelaire, or in Renaissance or pre-Raphaelite paintings — those are tasks for others better qualified than I. We've seen the fundamental importance of smell in sex right across the animal kingdom, noting crucially that in humans, the direct line of action between sex smells and reactive behaviour has been broken. I want to show that a powerful theme running through Western

---

[114] Drobnik, J. (ed) 2006 *The Smell Culture* Reader. Berg, Oxford.

literature and art is firmly based on the role of smell in sex, and particularly on the role of axillae as the seat of personal scent. Art and literature are products of the social, political, and religious constraints of the time of their creation, and to understand the context in which the Western world produced some of its finest culture, we need to understand the changing manner in which societal attitudes to smell have changed in recent centuries.

*   *   *

During the 17th and 18th centuries in Europe, a number of significant social developments were occurring. The Renaissance was underway, patchily at first before gathering momentum, and several important cities were growing quickly in size. On the one hand, knowledge of science and medicine was developing apace, while on the other, squalid and overcrowded housing conditions yielded to the scourges of plague on previously unknown scales. The Greek scholar Hippocrates taught that disease was caused by bad smells. This idea remained very much the mainstay of medical thought until the 19th century, and the gradual acceptance of the germ theory. Hippocrates viewed the smells of dank, foetid places, such as marshes, graveyards, and abattoirs, as directly responsible for disease. Malaria gets its name from the Italian words for bad and air — 'mala' and 'aria' — a reference to the stinking marshes, from the smell of which the disease was thought to emanate. Ironically, the marshes *were* responsible for the disease, but only because mosquitoes need marshy places in which to lay their eggs. Sick rooms were fumigated with juniper, rosemary, and garlic, and those tending the sick were advised to hold pomanders close to their nose at all times, or to stitch small sachets filled with rue, marjoram, mint, lavender, and quince rind into their clothing. Plague doctors carried small incense braziers with them when they went on their rounds, and breathed air through beak-like masks concealing *potpourris* of fragrant herbs. Medicines were always aromatic, for the only way to rid a person of a smell-caused illness was to chase out the bad smell, and with it the malady, with good smells. Physic gardens, from which physicians gathered their *materia medica*, were fragrantly beautiful oases, amid the stench of medieval cities.

Alain Corbin's *The Foul and the Fragrant* examines how social sensitivities to smell have changed over the last 400 or so years, mainly in France, but by extension to the rest of western Europe. From the time of the Enlightenment, Paris was one of the most important capitals in Europe. As its importance grew in medieval times, so did its population. It became heavily overcrowded and quickly became overwhelmed with the piles of human excrement and animal waste that accreted everywhere in the streets and laneways. All too often stinking pools of excrement lay around for weeks on end. Putrid vapours were everywhere, and in parts of the city where animals were slaughtered and leather was cured, and around cemeteries where bodies lay around prior to burial, people lived their whole lives without ever taking a single breath of odourless air. Even the soil beneath the city stank.

Things started to change following the ousting of Charles X in 1830 by Louis-Philippe of the House of Orléans. A far more liberal thinker than his conservative predecessor, Louis-Philippe owed his support to the *nouveau bourgoisie* who knew much more about the stench of city life than did any member of the previous ruling class. Under Louis-Philippe's rule, social hygiene movements set about clearing out slums, installing sewerage systems to transport human waste away from where people lived, cleaning up practices at abattoirs and cemeteries and so on. The newly emerging middle class was redefining society's relationship with smells. Gradually, bad smells came to be seen as the *consequence* of disease rather than its cause, and squalor and filth were recognized as the breeding ground for infection.

* * *

During the early part of the Enlightenment, perfumes were used not to allure but to mask a lack of personal hygiene brought about by few baths and no running water. The moralists of the day argued that bathing was harmful anyway, as nakedness led to immorality and therefore should be indulged in as sparingly as possible. Clothes had to have the *appearance* of cleanliness, and one way to make this happen was to perfume them. Heavy perfumes made of musk, ambergris, and civet were widely used; floral extracts were restricted to the elite.

At the court of Louis XV during the 18th century, courtiers were expected to use a different perfume every day. Cakes of incense shaped as little birds, sometimes decorated with feathers and even set up in tiny gilded cages, were commonplace at his court. Such 'Cyprus Birds', as the cakes were known, emitted supposedly aphrodisiacal scents, based on labdanum — the fragrant resin of a Mediterranean rock rose. Cyprus Birds would have witnessed much courtly licentiousness! As the 18th century progressed and the nobility preferred the light balsamic fragrance of springtime meadows, musk and heavy animal scents were ridiculed on the basis that such indiscreet perfumes of the lower classes might cause discomfort to the upper. That musk and civet are employed by their unwilling donors for sexual attraction in their own species didn't escape the moralists' eyes either.

The English authorities of the 18th century passed a law in 1770 that prevented women from trapping men into marriage through the use of various artificial devices, including the heavy perfumes of the day. The Act provided:

> 'That all women, of whatever age, rank or profession or degree, whether virgins, maids, or widows, that shall from and after this Act impose upon, seduce and betray into matrimony any of His Majesty's subjects by the use of scents, paints, cosmetic washes, artificial teeth, false hair, iron stays, hoops, high-heeled shoes, or bolstered hips, shall incur the penalty against witchcraft, and the marriage ... shall be null and void.'[115]

Henry Havelock Ellis, a British psychologist who wrote a definitive early 20th century tome on sex,[116] observed that women didn't wear musk to hide their own natural smell, but to emphasize it. Ellis' theme was taken up by adherents of the new philosophy of 'osphresiology' — the study of odours and the sense of smell. Osphresiologists

---

[115] Thompson, C.J.S. 1927 *The Mystery and Lure of Perfume.* John Lane The Bodley Head, London.
[116] Ellis, H. 1906 Studies in the *Psychology of Sex.* Volume IV Sexual Selection in Man. FA Davis Co., Philadelphia.

noted that many women sought out the strongest scents, most resembling animal smells, for purposes of entrapment. *Les putains*, they tell us, used strong animal smells to advertise their wares and attract their clients. The sexual role of smell thus became the target for social and moral reformers who pressed for the use of only floral fragrances for disguising natural body odour, abhorring the use of heavy animal perfumes by women. As Corbin observed: 'Vegetable perfumes re-imposed their delicacy; their function was to dampen female impulses and to signal a new system of control.'[117] *Parfumiers* were charged with maintaining society's norms, though the cleverest of them created concoctions that aroused desire without the wearer losing her modesty.

*   *   *

As the modern age unfolded, nation states across Europe built their societies on strong social stratification. The ruling classes had agendas to pursue that required a compliant and obedient underclass, for how else could the industrial revolution make them wealthy? Just as smell was the great social divider, it was also seen as having the ability to release pent-up animal desires in humans, potentially destroying carefully maintained social order. Sexuality was regarded as underpinning many social problems. During the 19th century physicians and judges focused their attention on the problem, and on the many faces of what was called 'sexual deviance'. In 1886 the Austrian physician Richard von Krafft-Ebing wrote a definitive work on human sexuality, aimed at a readership of judges and doctors to help them determine whether a person up on a charge before them could be regarded as a deviant, and therefore punishable by clearly prescribed laws. Considered unsuitable to be read by the public, he chose as his title *Psychopathia Sexualis: eine Klinisch-Forensische Studie*, writing some parts in Latin to put off all but the most persistent general reader, and using as inaccessible a style as the sometimes impenetrable

---

[117] Corbin, p 230.

19th century German language would allow.[118] Significantly, he defined all sexual relations, other than those occurring between strictly heterosexual couples, to some category or other of sexual deviance. Admission of pleasure gained from human body scent was enough to indicate deviance, and doubtless earn a heavy sentence.

Over in England, Havelock Ellis devoted over 130 pages of his *opus magnum* to smell. He reviewed, meticulously, everything that had been written about human smell published in the 18th and 19th centuries, much of it by physicians reporting clinical cases of patients with unusual responses to smells, and by expeditions reporting cultural uses of smell in exotic parts of the world. Ellis was in no doubt about the importance of the axilla as the site of human scent and its sexual significance:[119]

'When we are dealing with the sexual significance of personal odours in man there is at the outset an important difference to be noticed in comparison with the lower mammals. Not only is the significance of odour altogether very much less, but the focus of olfactory attractiveness has been displaced. The centre of olfactory attractiveness is not, as usually among animals, in the sexual region, but is transferred to the upper part of the body. In this respect the sexual olfactory allurement in man resembles what we find in the sphere of vision, for neither the sexual organs of man nor of woman are usually beautiful in the eyes of the opposite sex, and their exhibition is not among us regarded as a necessary stage in courtship. The odour of the body, like its beauty, in so far as it can be regarded as a possible sexual allurement, has in the course of development been transferred to the upper parts. The careful concealment of the sexual region has doubtless favoured this transfer. It has thus happened that when personal odour acts as a sexual allurement it is the armpit, in any case normally the chief focus of odour in the body, which mainly comes into play, together with the skin and the hair.'

---

[118] Krafft-Ebing, R F von. 1967 *Psychopathia sexualis*, with especial reference to the antipathetic sexual response. (Translation of 12th edition by F.L. Flaf) Mayflower-Dell, London.
[119] Ellis, H. 1906 *Studies in the Psychology of Sex* Volume 4 Smell, Chapter 3.

Ellis notes that the power of axillary scent witnesses itself only after a person has experienced some degree of arousal, and gives many examples from the literature and from his own observations. He speaks of the 'latent possibilities of sexual allurement by olfaction … inherited from our animal ancestors, still … ready to be called into play'. He regarded unusual interest in natural body smells as a deviance seen in people with personalities identified as 'olfactive types' which, he notes, were common among writers such as Baudelaire, Huysmans, and Zola. He also reports such interests in those whom he classified as 'olfactory psychic types' — people who become sexually aroused by body smells, known as *'renifleurs'*. What these European writers of the 19th century were telling us was that body smells have the potential to upset social order, and the challenges to aesthetic sensitivities brought about by the release of pent-up animal desires had no place in a manicured social environment.

*   *   *

Sigmund Freud was well acquainted with the works of Krafft-Ebing and Ellis, and paid some attention to the sense of smell in his writings.[120] He held convoluted views about smell, tracking clinical cases in which a patient had hang-ups about his or her sense of smell back to the patient's early relationship with his or her mother. Freud's view, that men *looked* while women *smelled,* was firmly based in the sexist cultural ideology of the *fin-de-siècle*, rather than on any biological evidence. Freud's position was, in part, built on what he saw as the uncomfortable fact that the scent of the human body had a strongly erotic quality.[121] In a world then still plagued by poor hygiene, he regarded the universal presence of arousing scents as an

---

[120] Freud, S. 2002 *Civilization and Its Discontents*. Penguin, London. (Originally published as *Das Unbehagen in der Kultur* 1929 ('The Uneasiness in Culture').)

[121] Freud dismissed the nose in his theory of psychoanalysis, despite having previously been captured by the notion of a naso-genital relationship posited by his colleague Dr Wilhelm Fliess. After falling out with Fliess over a plagiarism issue, Freud had remarkably little more to say about the topic. For much of his life he suffered nasal congestion, suppuration and sinus pain, and was treated with cocaine and surgery by Fliess. By all accounts, he had a poor sense of smell.

ominous warning for the maintenance of social harmony. Smell was too closely linked with animal sexuality for a harmonious civilization to exist, for it could so easily release visceral and libidinal behaviour. In Freud's view, individuals' reactions to scent encapsulated one of the greatest paradoxes confronting civilization; smell was placed right at the intersection of animal desire and societal shame, creating tensions that led to all manner of psychological repressions.

Smell being so closely linked to sexuality, Freud speculated that some process of 'organic repression' had diminished the sense of smell; he was a century and a half too early to know that his so-called 'organic repression' was actually the loss of VNO functionality. He noted the adoption of an upright posture, which raised the nose several feet above the ground, as the driving force behind the supremacy of sight over smell, and as a major factor in the 'organic repression' of the sense of smell. Accompanying bipedalism was the full exposure of the genital organs that had previously been hidden, leading, in his view, to shame and neuroses — though the basis of his determination isn't clear. He argued in *Civilization and Its Discontents* that society had many laws and rules restricting what citizens can and can't do, many of which run counter to immutable instincts, such as unbridled sexual desire and aggression towards sexual competitors. Between the tensions of these opposing forces lay discontent. The situation was not without a personal silver lining however; he and others established booming psychotherapy clinics to treat people suffering from what was seen as the inevitable endgame of the tension — the repression of instinct and sexuality.

Freud's writings show he regarded the nose, and what it could perceive, as lying at the heart of many neuroses, and when human body smell was involved, the problem was, by definition, sexual. All ailments became swept up in Freud's theory; we saw before how his colleague Wilhelm Fliess linked menstrual problems with what he thought were problems concerning the oversensitivity of the nose, relievable only by invasive nasal therapy. Underlying many of the therapies for the relief of symptoms concerning sexual physiology and behaviour was the view that an oversensitive nose was a throwback to our animal past. Until the nose had been desensitized the patient would remain in a kind of evolutionary no man's land; not animal, but not quite fully human either. Fortunately these ideas are no longer in vogue, and although civilization continues to be plagued with more discontents than can be counted, body smells don't rank amongst them.

European art was placed amongst these powerful social tensions, aided and abetted by God's domination over humankind, men's domination over women, and the domination of those with money over those without. The dominant philosophy about how smells were perceived, during the five centuries from the later Middle Ages to the start of the 18th century, was based on the sketchy understanding of the brain's anatomy provided by Galen's and Hippocrates' view of blood and the humours it contained. The Middle Ages were also a period when visual images came to ascendance as never before, helped by woodcuts and printing presses.[122] The senses were thought to report to ventricles of the brain, from where their stored products could be used to analyse the world. Renaissance artists were fascinated with the senses, and went to some lengths to communicate how they worked. The fact that the nose was inside the head, and very close to the brain, gave it a special importance; after all, the Bible tells us that God breathed life into Adam through the nose.[123] Beautiful smells were visualized through depictions of flowers and petals; bad smells through grim-faced people holding their noses. Roses figured extensively, with petals strewn around and blown on the wind.

As early as the 16th century, painters were wise to the power that olfactory allusions could bring to their art. Titian's famous work *The Venus of Urbano* (1538) invites the observer to mix the scent of the rose with the expectation of venery, in this most provocative painting. The canvas shows a nude woman reclining on a bed looking directly into the artist's eyes. In her right hand she holds a rosary

---

[122] Quiviger, F. 2010 *The Sensory World of Italian Renaissance Art*. Reaktion Books, London.

[123] Genesis Chapter 2: v 7 'And the Lord God formed Man of the dust of the ground, and breathed into his nostrils the breath of life; and man became a living soul.' King James version.

of flowers; her left hand rests in her pudenda as if to protect her modesty. It clearly isn't, because the artist could easily have painted a judiciously positioned fold of bed sheet, had her modesty been his intention. The woman's hand draws the observer's attention to the real message of the painting, which is a strongly scented one. She lies on a white sheet, but where this is rolled back the observer can see the bed's dark red mattress is decorated with roses, adding a further scented dimension to the visual image. In analysing this painting within the context of how the senses were depicted in Renaissance art, François Quiviger notes that these associations have not previously been recognized by art historians. In the two centuries after Titian, artists found ways to get around the increasingly oppressive social and religious tensions, and they did this by frequently placing an exposed armpit at the very centre of their compositions, so giving olfactory life to their otherwise sterile subjects, in a style of painting that continues to the present day.

Chapter *11*

Art and Literature

The complicated social and political milieu of the 18th and 19th centuries places the cultural context of art in immense complexity, which must be understood to appreciate it fully. Painters depicted religious scenes for display in churches, along with family portraits, war scenes, landscapes, still-life compositions, and paintings of children, horses, cattle, and even domestic pets. At the same time, another genre of painting featured the dalliances of gods, goddesses, and nymphs from classical Greek literature. The central subjects in these paintings were nude women, and the reason why women were so depicted isn't hard to understand. The great painters of the pre-Renaissance, and later, were nearly all men who generally painted to commissions from aristocratic landowners and titled noblemen — men with money. With men as the primary commissioners and consumers of art, it's hardly surprising that pictures of female nudes were commonplace. Most paintings depicted scenes implying some degree of male sexual power and domination.

The nude has been thoroughly analysed by art historians, including by Lord Kenneth Clark whose book *The Nude: A Study in Ideal Form* is an authoritative text on the matter.[124] The nude as a conceptual and

---

[124] Clark, K. 1956 *The Nude: A Study in Ideal Form*. Pantheon Books, New York. Reproduced with the permission of The Estate of Kenneth Clark, c/o The Hanbury Agency Ltd, 28 Moreton Street, London SW1V 2PE. All rights reserved.

artistic category, is very different from a painting of a naked person, as Clark notes:

'The English language, with its elaborate generosity, distinguishes between the naked and the nude. To be naked is to be deprived of our clothes, and the word implies some of the embarrassment most of us feel in that condition. The word "nude", on the other hand, carries, in educated usage, no uncomfortable overtone. The vague image it projects into the mind is not of a huddled and defenseless body, but of a balanced, prosperous, and confident body: the body re-formed. In fact, the word was forced into our vocabulary by critics of the early 18th century to persuade the artless islanders [of the UK] that, in countries where painting and sculpture were practiced and valued as they should be, the naked human body was the central subject of art.'

He goes on to consider the erotic in nude painting:

'In the volatile Victorian culture, artists and critics were forced to find a way to separate beauty from sexuality. Their initial attempts to cast out sexuality completely, resulted in highly repressed works, but to the censor's chagrin, audiences still found a way to project a sense of the erotic onto even the tamest subjects. Erotic art's subjective quality allows any work to develop a sexual charge. It is the viewer and his response to a work which produces the erotic.'

In the 19th century, nude paintings of nymphs and goddesses were depicted in poses first seen in classical Greek sculptures of male athletes, where the subject's weight is taken on one leg so raising that hip, and allowing the knee of the other leg to be bent slightly. Even if the model were reclining, as was often the case, the body's frame gives an impression of dynamic relaxation. The concept of the classical nude was an art form in itself, and not a subject of art. The subject was flawless, often painted in skin tones of marble to resemble the Greek sculptures and, in an olfactory sense, completely detached from reality. The skin had to be smooth and unblemished giving not the slightest suggestion of being sweaty or, worse, having any smell. The body had to be deodorized, and apart from that on the head, every trace of body hair had to be removed. The essential characteristic of the nude is that it's scrubbed clean of all traces of its animal origins, and presented as if it has little to do with the subject from which it was painted. In short, it had to be more perfect than evolution intended.

'Art', notes Clark, 'completes what nature cannot bring to a finish. The artist gives us knowledge of nature's unrealized ends'. To a scientist trained in classical zoology, this is a challenging concept suggesting that art and nature, at least on this matter, cannot be reconciled. What is much more understandable to a zoologist is the subtle way in which artistic imagery has managed to stimulate ancient neural pathways in the brain, which evolved when smell ruled the lives of our distant ancestors. By the use of this device some reality was injected into their artificial and unreal paintings. Just as the *parfumiers* at the court of Louis XV managed to create perfumes which subconsciously revealed some of that which they strove to hide, so by exposing the armpits of their models, painters of the 18th and 19th century injected a sense of the erotic into their work. This raises an important question about nature and art that appears to have escaped the attention of art historians.

Sculptures of Greek athletes seldom showed a subject's arm lifted above the head, unless the subject was throwing a javelin or the pose otherwise warranted it. In 18th and 19th century nude paintings, the exposed armpit quickly became a regular feature of the genre. What is most striking about such paintings is that the spectator's eye is drawn not to the nakedness of the subjects, nor to their exquisite flesh tones, nor even to what activity they are engaged in, but to their armpits, unclothed and exposed. The displayed armpit introduces a tension into a painting, triggering the imagination of an olfactory sensation. What the armpit has to offer becomes an element in the interaction between the painting and the spectator, even if it's shown shaved and scrubbed. Although raising an arm above the head can, and usually does contribute to a painting's poise and balance, it significantly restores animal identity to a model that's otherwise stripped of all reality. The armpit is a part of the body that has evolved under mutual sexual selection as an adornment designed for use in a privatized sexual context; its public display hints at an opening in a door that is normally kept shut. A displayed axilla in an otherwise sterilized depiction of a human body stimulates some of those 'latent possibilities of sexual allurement' noted by Ellis, imparting to a painting a significant olfactory, and human, dimension.

The subjects chosen for the genre of paintings depict mythical characters and include many Greek goddesses. Psyche and Venus among others, countless nymphs, and sometimes a few gods are regular subjects as their lovers. The titles of the pictures are misleadingly benign such as: *The Birth of Venus* (Alexandre Cabanel 1823–1889, *Reclining Venus* (Giorgio Giorgione 1477–1510), *Awakening of Psyche* (Guillaume Seignac 1870–1924), *Bacchante* (Jean-Baptiste Corot 1796–1875), and many more in a similar vein. Sometimes the artists push fantasy to one side and title their paintings for what they really are, such as: *The Temptress* (Dominique Papety 1815–1849) and *L'Abandon* (Guillaume Seignac 1870–1924). In some of these paintings — Seignac's *L'Abandon*, Bougeureau's *Birth of Venus*, and Corot's *Venus at her Bath* are excellent examples — the subject's face is either partly hidden, or declined in such a way as to direct the viewer's eye to the model's armpit.

Bacchanalian feasts were frequently painted showing wanton priestesses of the cult displaying their axillae; odalisques, satyrs, swarthy youths, and soft-skinned courtesans all engaged in frisky behaviours were common subjects. Even in biblical scenes, such as in *Toilet of Esther* (Theodore Chasserieu 1819–1856) and *The Temptation of St Anthony* (John Charles Dollman 1851–1934), the axillae are not disguised or covered, but the central point of interest. In John Roddam Spencer Stanhope's painting *Eve Tempted*, Eve's awkward posture as she reaches for an apple above her head while being tempted by the human-headed snake can only be explained by the artist's desire to expose her axilla to the viewer's gaze (Figure 11.1).

Carol Mavor suggests that when you look at a picture in this genre, the visual stimulus of your gaze releases memories of smells and their contexts, so linking the visual and olfactory senses together, denying supremacy of one over the other. So the paintings come to

Figure 11.1

(a) *Eve Tempted* John Roddam Spencer Stanhope circa 1877 Manchester Art Gallery.
(b) Advertisement for the Pearl Tobacco Company 1871.

life, giving more than just a gentle hint to the imagination.[125] Recall that the endocrinologist Aubrey Gorbman proposed that in the evolution of the vertebrate brain, the anterior pituitary transferred its scent-sensory function to the adjacent nervous system, so making its

---

[125] Mavor, C. 1998 Odor di femina. Though you may not see her you certainly can smell her. Cultural Studies 12 (1): 51–81.

stimulation possible by a range of sensory inputs additional to smell, including visual stimuli. In this case a visual input also triggers an olfactory imagination; a displayed axilla releases olfactory memory and both stimulate ancient neural pathways in the brain's limbic system.

European nude paintings of the 18th and 19th centuries provided fantastic relief from the social doldrums imposed by Christian morality, arranged marriages, and the many weighty strictures applying to European society, using the adjective 'fantastic' in its original sense of existing only in the imagination. From the Middle Ages onwards, Christianity subordinated humankind to the primacy of God, reducing an individual's ability to act instinctively. The portrayal of sexual power and control held by men over women, and the control held by God over humankind, lie at the centre of the murky orb in which these pictures were conceived, executed, and viewed. If we're shocked to find them so beautiful, it's because of their ability to stir subconscious neural pathways in our brains. To great effect, skilled artists expose the viewer to something encountered only during intimacy, making it voyeuristically respectable and available to anyone who cares to look. The use of mythical subjects, though painted from quintessentially human models, depersonalizes the paintings and opens them up for scrutiny without fuelling sexual aggression. We can sense the scene, yet we smell only canvas and turpentine.

While there are some examples of 19th century art in which the female axilla is shown unshaved, as in Courbet's *Femme á La Vague*, for example, such depictions are rare. Contemporary painters have changed the rules however, and now frequently paint the axilla unshaved. A striking example is Australian artist Brett Whiteley's surreal *Justine* (1986), one of a series of paintings of life on Sydney's Bondi Beach in the 1980s, that puts the unshaved axilla of his model — his wife Wendy — right at the dead centre of his contorted subject, who lies on the beach nonchalantly reading an eponymously titled book.[126]

---

[126] http://img.aasd.com.au/01450034.jpg.

Paintings displaying male axillae are much rarer than those displaying female axillae, and when they do the context is always non-sexual. One well-known example of male axillary display in a distinctly non-sexual context is the harrowing painting by Hans Holbein the Elder of the *Martyrdom of St Sebastian*, painted in 1815, in which a crossbow bolt pins the saint though his right axilla to the tree behind him.[127] Other bolts puncture the saint's neck, chest, side, and legs, but it's the one through the axilla that causes the spine to tingle.

*   *   *

There is another type of olfactory imagery in paintings that exerts an equally powerful, though more warming emotional response and it's the paintings dealing with soft femininity. This is something that earlier painters represented in usually rather staid and stiff formal portraits. John William Waterhouse was an English painter of the late Pre-Raphaelite school who focused on the relationship between the scent of flowers and femininity, with never an armpit in sight. Arguably his most beautiful work, *The Soul of the Rose*, painted in 1908, shows an elegantly dressed woman smelling a rose flower she draws towards her from a bush (see frontispiece). The woman is seen only in profile resulting in the focus of attention being her interaction with the rose and its aroma, and not her face. She seems intoxicated by the smell and has to steady herself with her left hand against the wall. She's quite oblivious that we are watching her. Had Waterhouse been able to paint the rose's real scent, his painting could hardly have carried a stronger message. In another of his works, *Gather ye Rosebuds While ye May*, two young women gather musk roses in an Arcadian landscape, leaning into one another as they work. We can't see the full face of either, but we're overcome with envy at the power of the flowers to command such attention from the young women.

Art historian Christina Bradstreet has analysed how painters of the mid-19th century depicted femininity and fragrance while noting

---

[127] http://m1.i.pbase.com/o6/93/329493/1/84412691.o64XrMJP.MUCDec06134.jpg.

a sublimed sexual theme hidden, but not extinguished, in their paintings.[128] She shows how the American painter Charles Courtney Curran (1861–1942) portrays femininity through beautiful real-life paintings of women and flowers, as well as through fantasy works in which fairy-women live among roses. Bradstreet holds that rose scent takes on a physical form in these fantasy works because Curran has captured it as the embodiment of the fairies, who need nothing more than it for their total existence. By fusing the perfume of flowers with the female form, femininity is delicately defined. But sex is never far from the surface, as her analysis of John Waterhouse's *The Shrine* (1895) shows. In this serenely beautiful picture, a young girl bends to smell a bowl of white roses; the observer voyeuristically catches her, as it were, in the vulnerable moment as she is seduced by the fragrance. In many of his paintings Waterhouse shows his mastery by painting subjects that completely shun the viewer, while at the same time demanding the same viewer to share their moment.

The flowers at the centre of so many 19th century paintings are roses. Analysis of rose scent shows enormous chemical complexity, with over 400 different compounds known. The main ingredients, however, are reminiscent of musk; old-fashioned simple roses lacking the multi-layered petals of the hybrid 'T' commercial flowers are called 'musk roses' for that very reason. Musk roses have been implicated in the work of Satan from the 16th and 17th centuries, with reports of people — even nuns — being possessed by the scent of the flowers. Supposedly it was the animalistic component of the rose's fragrance that attracted Satan, a component that imbues rose scent with its almost mystical powers.

Carl Linnaeus, the father of biological nomenclature, recognized musk as the most important plant fragrance for attracting insects for pollination, as reflected in the names attributed to many types of plants — musk roses, muscatels, musk thistles, musk orchids, and many more. In the animal world there are musk rats, musk oxen,

---

[128] Bradstreet, C. 2007 'Wicked with Roses. Nineteenth Century Art Worldwide vol 6 (1) Available at: http://www.19thcartworldwide.org/index.php/spring07/144-qwicked-with-rosesq-floral-femininity-and-the-erotics-of-scent.

musk turtles, musk lorikeets, musk duck, musk shrews, and of course, musk deer. Nineteenth century physicians regularly made mention of musk as the most restorative and invigorating of smells, perhaps reflecting the view of Dioscorides that the smell of musk, more than any other scent, exerts a profoundly erotic effect. Havelock Ellis identifies musk as 'the fragrance of sex', the universal aphrodisiac that pervades nature. This, then, is the smell that is recalled from the delicate paintings of fairies and flower pickers. Artists in the late 18th and 19th centuries knew nothing of the chemistry of rose scent, but their aesthetic awareness put onto canvas the ancient connection between musk and sex, that the animal in each of us instantly recognizes.

*   *   *

It's well known that sex sells. Advertisements may not be art in the strict sense, but they may share many elements of composition, form, and execution. There's hardly a commodity that's been spared advertisements grabbing our attention through the use of sexually suggestive images unrelated to the nature of the product itself. Once your attention has been grabbed, the role of the advertisement is finished; whether you buy the product ultimately depends on its utility and effectiveness, rather than the catchiness of its advertisement. One of the earliest examples of sex appeal in marketing was an advertisement for the Pearl Tobacco company in 1871, showing a nymph-like young woman riding on the waves, reminiscent of many classical paintings of Venus emerging from the sea, e.g., Cabanel's *Birth of Venus*, or Botticelli's painting of the same name (Figure 11.1). Her lower body is swathed in folds of diaphanous silk while her torso remains bare. Her right arm is raised above her head displaying her armpit.

The fortunes of many of today's retailing companies revolve around a trend towards making their advertisements ever more explicit, while not completely sacrificing the last vestiges of taste. Clothing companies are particularly aggressive in this respect, seemingly doing whatever it takes to grab more attention and with it, a greater share of the money in peoples' pockets. Calvin Klein, Abercrombie and Fitch, and American Apparel are all major companies feverishly vying for our attention, with increasingly suggestive and arresting advertisements. Several have come under attack at one time or another for publishing advertisements that are just a bit too raunchy, and which push the boundaries of good taste just a bit too far. Many of the poses adopted by their models expose the armpit to clear view, doing little to better show off the garment, but much to increase the erotic content of the advertisement, just as we saw in classical paintings. This device to grab attention is well seen in advertisements for American Apparel; their online catalogues and advertisement archives provide many examples of where an exposed armpit introduces eroticism to an advertisement.[129]

Unsurprisingly, advertisements for women's apparel target women, as they are the buyers and wearers of female apparel, yet the advertisements have all the characteristics of the fantasy paintings of a century or two ago, commissioned by men for men as spectators. Unlike those classical images purportedly of mythical figures, in modern advertisements it is the model that is the object of attention, and not a proxy for a mythical subject as was formerly the case. Indeed, American Apparel displays a short biography of each model in their advertisements, focusing their advertisements more about the model than about the garment. Is it not curious that advertisements targeting women have the axillae exposed, as we saw in so many classical paintings? The fact that axillary organs evolved under the specialized circumstances of mutual sexual selection puts them squarely in the arena of *intra*-sexual competition. Displaying the source of one woman's individual scent signature releases a 'me too' response in other women of a similar age. In all these hidden biological imperatives, the little black dresses, or whatever the advertised article of apparel, are almost incidental, though the retailer's banking on the fact that you'll end up buying them.

*　　*　　*

In recent years there has been a surge in the number of books that deal with the sense of smell. Much of it goes back directly to Alain

---

[129] www.americanapparel.com.

Corbin's scholarly work. Patrick Süskind's gripping psychological thriller *Perfume*[130] was inspired by Corbin's descriptions of the ferocity of the stench on the streets in medieval Europe and I recommend you read it, if you haven't already done so. Not only does it accurately describe how early perfumes were made, it also graphically demonstrates the grip held by smell on the human psyche. Contemporary literature is challenging the long-held relationship between sweet smells, goodness and femininity, and bad smells, savagery, depravity, and badness, often through the eyes of women authors. In Jamaica Kincaid's 1996 novel *The Autobiography of my Mother*, for instance, the narrator's love of her own body odour is used to resist the perception she's unloved by others. Danuta Fjellestad, whose analysis of the aesthetics of smell in contemporary literature we've met before, shows us that the narrator draws comfort from her own body odour and regards floral scents as threatening. As Fjellestad says:[131]

> 'It needs to be stressed that the significance attributed to smell(s) is culture- and time-specific. And since odours are invested with cultural values, their cultural coding suggests models for marking and interpreting others as Others *[sic]* and for writing scripts of interaction between selves and others *[sic]*. Moreover, since the olfactory sense, like other senses, is manipulable and affected by our beliefs ... changing the beliefs about smell in our post-colonial world full of "others", a re-coding of smell may have far-reaching political and social consequences...'

An exactly opposite view was taken by the ruling classes in premodern Europe, who regarded the body's natural odour as threatening to orderly society. Modern writers are clearly changing the cognitive terrain occupied by smell.

Sentiments like those in Kincaid's novel were never expressed by writers of the past. To them, smell's grip on human emotions was

---

[130] Süskind, P. *Perfume. The Story of a Murder*. Trans. John E. Woods. Vintage International, New York.

[131] Fjellestad, D. 2001 Towards an aesthetic of smell, or, the foul and the fragrant in contemporary literature. CAUCE Revista de Filiogia y su Didáctica 24: 637–651.

solid, allowing the elicitation of non-challenging responses in the reader to be experienced. The odour of a loved one invariably gave rise to the most eloquent of prose, extolling desirability and susceptibility. Consider these lines by the English poet Robert Herrick, a 17th century Devonshire parson with an unrivalled interest in the body scents of his various lovers:

'Tell, if thou cans't, and truly, whence doth come
This camphire, storax, spikenard, galbanum;
These musks, these ambers and those other smells
Sweet as the vestry of the oracles.
I'll tell thee; while my Julia did unlace
Her silken bodice, but a breathing space:
The passive air such odour then assum'd
As when to Jove great Juno goes perfum'd.
Whose pure immortal body doth transmit
A scent, that fills both Heaven and earth with it.'[132]

In a similar vein, Homer[133] described Juno's meeting with Venus as an olfactory experience:

'Here first she bathes, and round her body pours
Soft oils of fragrance, and ambrosial showers.
The winds, perfumed, the balmy gale conveys
Through heaven, through earth, and all the aerial ways.
Spirit divine! Whose exhalation greets
The sense of gods with more than mortal sweets.'

The *Song of Solomon's* writer describes his love with evocative olfactory allusions:[134]

'How sweet is your love, my sister, my bride!
How much better is your love than wine,

---

[132] '*Upon Julia unlacing herself.*'
[133] Iliad XIV.
[134] Song of Solomon (King James version).

And the fragrance of your oils than any spice!
Your lips distil nectar, my bride;
Honey and milk are under your tongue,
The scent of your garments is like the scent of Lebanon.
A garden locked is my sister, my bride!
A garden locked, a fountain sealed.
Your shoots are an orchard of pomegranates
With all choicest fruits,
Henna and nard,
Nard and saffron, calamus and cinnamon,
With all trees of frankincense, myrrh and
With all chief spices.'

In the Pakistani writer Saadat Hasan Manto's short story *Bu* (*Smell*), the narrator recalls the memory of the scent of a *Ghatan* girl's armpits throughout his many subsequent conquests:

'It was like the fresh smell of earth sprinkled with water — but no, this odour was different. It was not an artificial smell like that of lavender or attar, but something natural and eternal, like the relationship that has existed between man and woman since the beginning of time.'[135]

The overpowering element of sexual attraction in these passages is that the very essence (the word itself depicts the fundamental tools of the *parfumier*) of the beloved is encapsulated in their fragrance. Their intimacy is frank to the point of being disturbing, as if the reader violates the subject's sanctity by treading too close to a secret and vulnerable zone. The librettists of Bizet's *Carmen* movingly exploited this concept when the young soldier, Don José, sings of the scent of a flower thrown to him by his lover Carmen; the flower, redolent with the smoky musk of Carmen's bosom falsely kept hope alive of a future with her as he languished in gaol.

---

[135] Saadat Hasan Manto '*Smell*' © The Annual of Urdu Studies No. 27 (2012); with permission of the editor.

In the days when soldiers on the battlefront received news from home only through letters and cards, handwritten pages carried the faint scent of the writer's hand. I'd like to think such faint reminders of the essence of loved ones and home similarly brought memory, hope, and comfort. There are reports that gum leaves were sometimes included in World War I letters sent to Australian soldiers on the faraway European front, so that the scent of the Australian bush would evoke pleasant memories of home. In the Australian War Memorial in Canberra there are several examples of 'perfume cards' — small rectangles of perfumed card put out by perfume houses to advertise their fragrances — sent by Australian troops in France in World War I to their loved ones back home. Had they not been away fighting a war, they'd likely have bought their wives and girlfriends *flaçons* of the real thing.

*   *   *

Cultural appreciation of the sense of smell has come a long way in the last three decades, elevating it from a sense of little importance, to one in which people wish to indulge with considerable enthusiasm. Jim Drobnick, a contemporary art historian at the Ontario College of Art and Design argues that, in the Western world at least, scents restore life to a world that has been over-sanitized, scrubbed so clean as to have lost much of its interest.[136] Art installations, in which patrons are more participants than viewers, go some way to restoring the broken link between people and their environment. The English writer Somerset Maugham felt much the same way, writing a perceptive essay linking the sanitization of the West with a collapse of true democracy.[137]

On a journey in rural China in the 1920s, Maugham had taken the best room at a small country inn when a high official arrived for the night, lambasting the poor innkeeper and everyone else in the

---

[136] Drobnick, J. 2000 Inhaling passions: art, sex and scent. Sexuality and Culture 4 (3): 37–56.

[137] Maugham, W.S. 1922 Democracy in: *On a Chinese screen*. George H Doran Co, New York. Reprinted with permission of United Artists on behalf of The Literary Fund.

vicinity, because he would have to make do with a mere servant's room. Maugham continues:

'An hour later I went into the yard to stretch my legs for five minutes before going to bed and somewhat to my surprise, I came upon the stout official, a little while ago so pompous and self-important, seated at a table in the front of the inn with the most ragged of my coolies. They were chatting amicably and the official quietly smoked a water pipe. He had made all that to-do to give himself face, but having achieved his object was satisfied, and feeling the need of conversation had accepted the company of any coolie without a thought of social distinction. His manner was perfectly cordial and there was in it no trace of condescension. The coolie talked with him on an equal footing. It seemed to me that this was true democracy. In the East, man is man's equal in a sense you find neither in Europe nor in America. Position and wealth put a man in a relation of superiority to another that is purely adventitious, and they are no bar to sociability.

When I lay in my bed I asked myself why in the despotic East there should be between men, an equality so much greater than in the free and democratic West, and was forced to the conclusion that the explanation must be sought in the cesspool. For in the West we are divided from our fellows by our sense of smell. The working man is our master, inclined to rule us with an iron hand, but it cannot be denied that he stinks: none can wonder at it, for a bath in the dawn when you have to hurry to your work before the factory bell rings is no pleasant thing, nor does heavy labour tend to sweetness; and you do not change your linen more than you can help when the week's washing must be done by a sharp-tongued wife. I do not blame the workingman because he stinks, but stink he does. It makes social intercourse difficult to persons of a sensitive nostril. The matutinal tub divides the classes more effectually than birth, wealth, or education. It is very significant that those novelists who have risen from the ranks of labour are apt to make it a symbol of class prejudice, and one of the most distinguished writers of our day always marks the rascals of his entertaining stories by the fact that they take a bath every morning. Now, the Chinese live all their lives in the proximity of very nasty smells. They do not notice them. Their nostrils are blunted to the odours that assail the Europeans and so they can move on an equal footing with the tiller of the soil, the coolie, and the artisan. I venture to think that the cesspool is more necessary to democracy than parliamentary institutions. The invention of the "sanitary convenience" has destroyed the sense of equality in men. It is responsible for class hatred much more than the monopoly of capital in the hands of the few.

It is a tragic thought that the first man who pulled the plug of a water closet with that negligent gesture rang the knell of democracy.'

The social and sanitarian reformist movements of the Enlightenment might not have agreed.

As the stinks and smells of medieval cities abated, the 19th and 20th centuries saw the emergence of generations of city dwellers who knew little of the smells familiar to their parents and grandparents. Surrogate scents gradually made their way into ordinary household commodities as consumerism took hold. For some household commodities, such as natural gas, a putrid smell was added for safety reasons so that unlit burners could be detected more easily. There's no reason, however, why washing-up detergent, furniture polish, or toilet paper should be made to smell of pine forests or lemon groves, but smell of them they do. The notion that the good smells good, and the bad smells bad is so deeply engrained in the human psyche that manufacturers of everyday products believe we won't buy their products unless they emit nice smells. Even household antiseptics and cleaning fluids are given a pleasant smell, apparently to help them ward off infection and disease, reminiscent of how *eau de cologne*, and other European plague waters, were supposed to keep the Black Death at bay, but failed miserably.

*   *   *

Constance Classen, David Howes, and Anthony Synnott in their book *Aroma*[138] have pointed out how the social history of smell, in the West at least, has been ignored in the wake of municipal and social sanitization of the past few centuries, taking the Western world into

---

[138] Classen, C., Howes, D., and Synnott, A. 1994 *Aroma*. Routledge, London.

a long period of what they call 'olfactory silence' — to use another apt, but sensorily inappropriate word. Smells have always been there, of course, but Western culture is only now catching up, and filling the silence. We can now see art installations in galleries which place smells at their centres, find books on the cultural history of smell, encounter retailers artificially scenting their premises to subliminally enrich your shopping experiences (code for enticing you to spend more money), read an avalanche of books on aromatic cooking, and spend huge amounts of money on runaway sales of perfumes and fragrant cosmetics. Smell has been re-calibrated in the modern culture.

The West's over-sanitized world has to be perfumed to give everything a nice smell, for anything that does not smell nice confronts our new sensitivities. Maybe Somerset Maugham was right; perhaps over-sanitization has weakened democracy.

*   *   *

There are many beautiful paintings of Venus, Psyche, and all kinds of Arcadian frolics in which no axillae are exposed, but there are many in which they are. Equally, there is a wealth of exquisite love poetry and literature making no mention of a beloved's smell, but there is much that does. Art is not made great by the artists' or writers' recourse to any one particular device, but by the complex appeal the work makes to the emotions as a whole, stimulating the viewer's or reader's experiences and background. As we don't all have the same personal experiences, and can't all draw upon the same well of emotional witness, we each respond to art differently, each in our own way. Binding us all together however, irrespective of age, race, religion, social standing, wealth, education, and everything else that makes us different from our fellows, is that we've all shared the same evolutionary olfactory path, and possess the same neural circuits with which our brains give meaning to smell. Each one of the seven billion of us took our first breath through a soup of smells — of the fluids that nurtured our first nine months, of blood, sweat, milk, armpits, urine, faeces, and the smell of apocrine glands. Every time we fed from our mothers' breasts we breathed the scent of life and it became imprinted in our memories. The power of smell in paintings, in poetry, literature, and advertisements appeals to the deeply hidden brain pathways that have lacked functional stimulation for millions of years, activating a part of our psyches concealed only by the flimsiest tissue of social veneer, like an ice bridge over a crevasse, subconsciously alerting us to a dream world not known to modern humankind. That dream world wasn't known to our recent ancestors either, but only to life before 'ADAM' came along to set us free from the behavioural slavery that hallmarks the animal kingdom.

Chapter 12

Adam's Nose and the Making of Humankind

Our investigation into the evolution of Adam's and Eve's noses has taken us from the first appearance of life on Earth, to the most highly evolved and successful species ever to walk on its surface. On the way we've seen how a few of the inestimable number of swimmers, fliers, creepers, and crawlers with which we share the Earth sniff out their food and their mates, living just about every aspect of their lives through smelling organs arranged in a vast number of shapes and forms. A universal use for the sense of smell in animals is in the coordination of sex; we'd struggle to identify any species that doesn't use its smelling system for this purpose. It's unfortunate that humankind perceives and understands the world visually and acoustically, though it's understandable why this is so. Olfactory components are ascribed minor significance in our mental constructs of the environment, because they play such a small conscious part in our rational lives. This explains why the similarity of manner in which sex is coordinated in us and in other species, which have been around since the Cambrian epoch, comes as such a surprise. We'd like to think we've risen above brute creation but, at least in this regard, we haven't. If we were asked to reflect on the human animal in terms of its sense of smell, our answer might go something like this.

Adam and Eve don't use their noses very much in day-to-day survival; eyes and ears have taken over watchdog functions that remain firmly the province of the sense of smell in most of the animal kingdom. While their noses are far from the best in the animal kingdom, they are nonetheless able to discriminate between many more smells than their eyes are able to define hues, or their ears to distinguish tones. Their simple nostrils and nasal passages are no match for a dog's, and mostly they can't follow a scent trail. They have half as many functional smelling genes as dogs but, to compensate, devote much more brain power to smell processing. Their noses are employed subtly to challenge the immune systems of potential mates, for which purpose both Adam and Eve are equipped with specialized axillary scent organs designed to promulgate subconscious clues about them. Eve's are slightly better endowed with apocrine glands than Adam's, but both are well developed. Eve has a weaker, or less intense smell than Adam, possibly because Adam's contains more compounds derived from testosterone than Eve's. The sexual orientation of both influences not only a smell's characteristics, but also how it's perceived.

Fragrances and aromas have played a central role in human cultural development from the earliest of times. Altars of Zoroaster, Confucius, and the temples of Memphis and Jerusalem were no strangers to billowing fragrance. We now know that incense smoke isn't as innocent as it seems; an essential ingredient in frankincense smoke alters the mind, inducing a sense of well-being. Ancient texts set out in great detail how sweet-smelling resins and gums are to be offered, by whom, when, and for what purpose. They tell us that incense was to be kept holy, and not used to perfume the body, and that those breaking the rule would be shunned by the church and cut free from the Deity's agenda for them. Anointing oils, made from the finest perfumes were, and still are, widely used in worship and coronations, where they symbolically link a human monarch's power to rule with sanction by the incorporeal gods. But in spite of repeated admonishments to the contrary, early bodies were perfumed then, as now, with the finest fragrances available. None other of Earth's 7.7 million kinds of animals applies perfume to their bodies for the purposes of inducing pleasure as we do. Fragrant concoctions have been made for millennia, with the best of them based on the sexual attractant smells of musk deer, civet, beaver, or plant resins and juices smelling like them. We don't allow ourselves to smell of the animal we are, yet we have an atavistic urge to smell of animals.

Generally speaking, Adam and Eve don't much like the smell of their fellows — at least on casual acquaintance. Eve has a better sense of smell than Adam, though her acuity waxes and wanes as her monthly cycle progresses. Adam prefers the smell of Eve as she enters the fertile phase of her monthly cycle over her smell during the infertile phase. As she herself approaches ovulation, Eve prefers scents of Adams that have the greatest degree of bodily symmetry to the scents of less symmetrical Adams. During the infertile phase of her cycle, however, she shows no preference. In monogamous couples, Eve's scent during the fertile phase of her cycle elevates Adam's blood testosterone level. As she enters her infertile period, Adam's testosterone level declines. Mothers and fathers can identify their offspring by their smell, and offspring can readily distinguish their mother's breast from the breasts of other lactating mothers.

As Adam and Eve grow up and become sexually mature, their ability to perceive smells associated with sex hormones begins to diverge. Eve perceives the scent of testosterone derivatives in Adam's smell with undiminished acuity, becoming even more sensitive to it at around the time she ovulates, while Adam's ability to perceive them declines, even to the point of extinction in some young men. Although there is an abundance of evidence in mammals that sexual development is influenced, if not actually driven by smells passing between mother and offspring, and between siblings, the existence of such a phenomenon in humans has not been subject to much investigation.

This brief reflection on humankind's nose leaves more questions unanswered than answered. In particular, while human sex is coordinated by mechanisms that can be traced right back through evolutionary time, we simply don't know if smells are in any way as important to us as we grow up as they are to animals. The evidence from animal studies about how the smell environment of birth and infancy guides the course of later sexual maturation is so compelling that questions need to be asked about it in our own species. Here again though, we find that the domination of our visual conceptualization of the world blinds us to everything else. Questions about the smells of parents and families haven't been asked, so we have no answers. At a time when modern family structure is changing faster than ever before, aided and abetted by science's lifting of the constraints of biological reproduction imposed by evolution, our lack of knowledge is verging on the dangerous.

*   *   *

Following a radio broadcast I made many years ago on the sense of smell, I was contacted by a woman listener in her mid-thirties whose mother had offered her for adoption soon after birth. She had recently tracked down her biological mother and on first meeting was struck by her smell. She knew in an instant that the woman was truly her mother and couldn't have been anyone else. The smell appeared to linger in the daughter's nostrils for days after their short meetings. A lifetime's feelings of rejection fell away when the daughter smelled her mother's scent, and a sense of contentment overwhelmed her.

Something special happens in our olfactory lives the moment we are born. We become imprinted on the scent of our mother's body, recognizing the scent of her breast and milk. Breastfeeding, and bottle feeding too, if administered by the birth mother, brings the infant's nose close to its mother's axilla and its scent, and that too, becomes imprinted. If we are at all like animals, early scent exposure can be expected to play a part later in life when the time comes for us to choose a mate. This probably expresses itself in the children of natural parents choosing partners whose immune system configurations are compatible with their own, while children deprived of parental scent may more frequently choose partners with immune system configurations not sufficiently different from their own. This is speculative, of course, as no longitudinal studies of these matters have been conducted.

Family arrangements have been shown to influence the time at which girls first start their periods, called the age of menarche. Although it hasn't been shown that family smells are the cause, knowledge of what happens in the animal kingdom should ring some bells. Girls growing up in families where the father is present have later menarche than girls in families where the father is absent. In a wide-ranging study, Robert Matchock and Elizabeth Susman surveyed almost 2,000 female college students, asking them to complete questionnaires about the composition of their families, whether they grew up in the country or in a city, and at what age they started their periods.[139] Three factors were associated with earlier than average age of menarche: absence of a girl's biological father, the presence of half-and step-brothers in the household, and growing up in an urban environment. Absence of a girl's biological father accounted for advancement of menarche by about three months. This may seem insignificant in today's world when the average age of first pregnancy is currently over 25 years (in the USA), but when our species was evolving, and life expectancy was much lower than it is today, three months delay in the start of a woman's reproductive life would have had a measurable effect on her lifetime's reproductive output. Also correlated with a delay in the age of menarche was the presence of a subject's sisters in the household, and was particularly obvious when there were older sisters. In this respect, we are a lot more like mice than we realize.

*　　*　　*

There's an outstanding account in the literature of a young boy, for whom the smell of his parents appeared to be important in his development. Michael Kalogerakis, a New York psychologist, followed the development of the child, who had an unusually acute sense of smell, from when he was two years old until he was about five.[140] Until the child had passed his third birthday, he seemed at ease with the smell of both his mother and father, but before he turned four he started to reject his mother's smell when he crawled into his parents' bed in the early mornings. This soon passed and the boy became relaxed, seeking the close bodily contact normally occurring between mothers and their young sons. When he was four, he started to find the smell of his father, and particularly his father's armpits, disagreeable.

---

[139] Matchock, R., and Susman, E.J. 2006 Family composition and menarcheal age: anti-inbreeding strategies. American Journal of Human Biology 18: 481–491.
[140] Kalogerakis, M.G. 1963 Role of olfaction in sexual development. Psychosomatic Medicine 25: 420–432.

Kalogerakis reports that this response was specific to his father and not to any man; the boy showed no antipathy towards his uncle, who was a regular visitor to the house. The observations ceased when the boy was a little over five years of age, and nothing more is reported about his eventual passage through adolescence into adulthood.

While this case may be unusual, and certainly was insofar as the child had an exceptional olfactory ability and was clearly very articulate, it indicates that the child's early olfactory environment was not a passive element in his life. Although there's little literature on this subject, psychiatrists agree that children find parental body smells comforting until they are aged about four or five, after which time they progressively find them unpleasant, or even repulsive. It's argued that the change in pleasantness helps to avoid incest and promotes exogamy.[141] Studies on children raised by their biological parents beg the question whether children brought up in family arrangements different from that within which our species evolved, show latent effects of social odour deprivation.

I've argued before that our pre-human ancestors evolved in broadly monogamous families on the basis of anatomical features; (lack of conspicuous male-only adornments, and lack of marked sexual dimorphism), both traits being associated with monogamous mating systems. Twenty-first century humankind lives in a variety of family arrangements and is by no means universally monogamous; the American Community Survey for 2010 shows there are over 590,000 same-sex households in the USA (about 0.5% of the total), with about 20% of them having children, equally divided between male and female same-sex couples.[142] While the proportion of children living in non-traditional households is small, the consequences of this modern social phenomenon are yet to work their way through society.

The current Western-world debate about same-sex, transgender, and intersex marriages, and the right of couples of all orientations to

---

[141] Buxbaum, M. 1983 Olfaction and sexuality. The Jefferson Journal of Psychiatry 1(1) Available at: http://jdc.jefferson.edu/jeffpsychiatry/vol1/iss1/3.
[142] US Census Bureau 2010 American Community Survey. Available at: www.census.gov/acs/www/.

have children, draws fierce views on all sides. Leaving aside the torrents of swirling religious and moral invective, medical science's ability to overcome the evolutionary dictates of biological reproduction has placed the rights of adults equally alongside those of children. The whole purpose of reproduction — recalling the principle of the selfish gene — is to give the offspring the best chance of carrying the parents' genes through to the next generation. As Bertrand Russell so aptly remarked: '…it is through children alone that sexual relations become of importance to society and worthy to be taken cognizance of by a legal institution'.[143] The issue for Adam and Eve is whether the consequences of being brought up in a household where there is only the scent of one parent, or two people of the same sex, only one of whom — or neither — is a biological parent, is likely to show up in later life. When a child is being imprinted on familial scents, the absence of the smell of one or more of its parents may turn out later to influence his or her psychosexual development. Sociologists report that the consequences of family disruption show up only when relationships become the focus of a young person's life, and not so much before,[144] and the same may turn out to be true for scent disruption. We simply don't know if the odour environment of same-sex households will have any effect on children raised in them, though knowledge of animal studies suggests the issue needs examination.

*   *   *

There's another area of human experience that shows childhood smells make lasting impressions. When Marcel Proust famously dunked a madeleine biscuit into a cup of tea, he found the smell wafting upwards transported him back in time to when he was a child sitting in his grandmother's kitchen. The smell went further, reminding him of people, places, and events from long ago. Proust was no different from you and me; you'll be as familiar as I with the phenomenon of suddenly being taken back to your first classroom, or a long-forgotten

[143] Russell, B. 1929 *Marriage and Morals*. George Allen & Unwin, London.

[144] Wallerstein, J., Lewis, J.M., and Blakeslee, S. 2001 *The Unexpected Legacy of Divorce*. Hyperion, New York.

village shop by the whiff of some smell randomly encountered as you walk along. What usually surprises is the strength of the memory recall; transport back through the decades happens instantaneously.

When we're young, and to a lesser extent as we grow older, smells imprint themselves into our brains. The phenomenon is well known in mammals, though poorly understood. It starts with the first breath a newborn takes. In species in which many young are born over a short time period, as happens in those that breed during a short and intense mating season, olfactory imprinting serves the purpose of enabling the mother to recognize her own young from among the many. The imprinted bond between mother and young is very strong; shepherds have trouble persuading a ewe whose lamb has died to foster-feed an orphaned lamb, generally only succeeding by skinning the dead lamb and wrapping the skin around the orphan, such that the orphan now smells of the ewe's dead lamb. Baby rabbits are imprinted on the scent of their mothers' nipples and use the scent clue to waste no time finding a nipple when the mother returns to the burrow to feed them. In humans too, newborn babies show a stronger response to the scent of their own mother's breast than the breast of another woman of the same lactational age. Although a ewe will actively butt away an orphaned lamb that approaches her for milk, the same doesn't happen in humans because we have the advantage of rational intellect. Wet nurses have been used since ancient times, particularly by the ruling classes, and still are in certain parts of the world.

Smells experienced early in a rat's life have been shown to have a life-long effect. Many years ago Thomas Fillion and Elliott Blass from Johns Hopkins University in Baltimore famously treated the nipples and bellies of female rats that had newly given birth with a citrus perfume, and observed the effect on their litters in a series of groundbreaking experiments on smell and sex.[145] They kept up the perfume treatment until the young rats were weaned, whereupon they were separated from their mothers. When they became sexually

---

[145] Fillion, T.J., and Blass, E.M. 1986 Infantile experience with suckling odors determines adult sexual behavior in male rats. *Science* 231: 729–731.

mature, at about three months of age, and introduced to female rats that smelled either of citrus or smelled naturally, the males immediately chose the citrus-smelling females over the others and mated quickly. They didn't completely shun normal-smelling females, but showed a strong preference for those smelling like their mothers. What the study showed was that the scent environment of infancy plays a strong role in the rats' sex lives later in life.

*　　*　　*

Leaving the topic of sex and smell to one side, it's now becoming apparent that many types of psychiatric diseases are associated with a reduction in a person's ability to detect smells at low concentrations, or to identify common smells.[146] People with dementia, schizophrenia, and a suite of major depressive disorders have olfactory bulbs and piriform cortices of reduced size; in dementias of the Alzheimer's type, all parts of the brain that process smells are affected. The overlap in function between emotional and olfactory brain structures, and in particular the limbic system and the orbito-frontal complex, provides the basis for the association, while also providing an opportunity for the development of non-invasive, diagnostic tools.

A new area of research is suggesting a relationship exists between certain personality types and olfactory ability. People who score more highly on an accepted psychopathy scale (people who more strongly agree with statements such as 'I purposely flatter people to get them on side', 'People sometimes say I'm cold-hearted', or 'I never feel guilty hurting others') had more difficulty discriminating between a set of well-known smells than non-psychopaths.[147] The orbito-frontal complex of the brain seems to be involved — the part that constrains emotions and imposes rationality. People with psychopathic traits can

---

[146] Atanasova, B., Graux, J., El Hage, W., et al. 2008 Olfaction: a potential cognitive marker of psychiatric disorders. Neurosceince and Biobehavioral Reviews 32: 1315–1325.

[147] Bettison, T.M., Mahmut, M.K., and Stevenson, R.J. 2013 The relationship between psychopathy and olfactory tasks sensitive to orbitofrontal cortex function in a non-criminal student sample. Chemosensory Perception 6: 189–210.

detect smells at the same concentrations as others; what they have trouble with is making sense of them. This new area of research can be expected to deliver deeper understanding about how the sense of smell interacts with personality type, and may help with better diagnoses.

*   *   *

Our personalities guide our likes and dislikes just as much as rational evaluation of a product, at least when it comes to the things that marketers want us to buy. Traditionally, marketers have happily advertised household products with reference to pine, lemons, forests, and the many smells of freshness, and are now leading us to link fragrance to well-known brands with no traditional relationship with fragrance. For example, manufacturers of cars have traditionally emphasized power, comfort, fuel consumption, and other things that rightly belong in the world of motoring. Recently however, they've been using the brand power of their vehicles to market scents. Jaguar — builders of up-market cars for almost a century — has announced several branded scents, including one specifically for women's use in an interesting extension of the brand's traditional appeal to men. Hummer, on the other hand, has remained true to the brand's original image. Its *Fragrance for Men* is full of sandalwood, cardamom, musk, and other macho smells; no version for women is anticipated any time soon.

Around the world, hotels are waking up to the marketing power of fragrances released under their own house brands, rather as perfume houses have long exploited the brand of celebrities to sell their products.[148] *Le Sirenuse* in Positano has created *Eau d'Italie*, used in all the hotel's soaps, shower gel, shampoo, and restroom toiletries, and sprayed around the lobby and fitness centre by enthusiastic staff. Incense, bergamot, blackcurrant buds, lichen, cedar, yellow sweet clover, and musk combine to give you the smell of warm terracotta, very appropriate for that most beautiful part of Italy. *Le Bristol* in

---

[148] Qantas Travel Insider. 'Hotel scents: How bespoke fragrance affects your stay.' April 2013 Available at: http://travelinsider.qantas.com.au/hotel_scents_accommodation_where_to_stay_room_fragrances.htm.

Paris has had *Le Bristol Paris* created for it by a leading French *parfumier*. The fragrance imparts a fresh floral aroma combining the scent of cut grass, freesia, lily of the valley, and white rose with sandalwood, cedar, and leather, all inspired by 'the hotel's fresh green garden as well as its numerous floral arrangements and cosy period furniture'. It's kept out of guest rooms but you'll encounter it in corridors and the hotel's public areas. The prize for hotel perfumes must surely go to Buenos Aires' *Faena Hotel* whose house perfume, *Fragrancia Catedral*, has captured the incense-rich aroma of cathedrals. Based on oriental woody fragrances, *Fragrancia Catedral* takes its inspiration from 'Walking in the woods in childhood; the idea of a tree as shelter; the nurturing scent of an elderly pine tree.' It's in the corridors and linen cupboards, and you probably won't need reminding that pine tree resin smells of warm, musky animal bodies. Expect the marketing of fragrances linked to commodity brands to continue apace as marketers become increasingly savvy about the power scent has over us; expect the range of signature scents to be limited only by the marketer's imagination and the methods of delivery to take you into unexpected places.

The retailing industry's also alert to the association between brand and smell. Visual similarity across all branches of large retail stores is commonplace; you know from the *décor* and design exactly which retail chain store you are in, irrespective of which city you are visiting. The 'scent of place' is something we all respond to; the smell of Woolworth's department store in the small town in England where I grew up is seared into my memory. I believe it came from the wooden floors, or whatever was put on them to keep them looking good, rather than some scent introduced into the atmosphere, but whatever it was it's remained with me throughout life. Expect to find that branches of major stores have the same smell in future. The smell will probably be one of universal attraction, such as vanilla or clove; it won't be musky or have animal overtones. The scent will be designed to make you feel comfortable with the store's ambience, inducing you to linger a bit longer for, as their research shows, the longer you linger, the more you'll spend.

Learning that over 100 billion lids for take-away coffee cups are consumed annually prompted an enterprising Hong Kong business to design a patch of impregnated plastic that emits a strong coffee smell, which can be attached to the lid during manufacture. Warmth from the real stuff below heats it up, filling your nostrils with an enhanced coffee experience that possibly encourages passers-by to buy a cup. Some bicycle helmets now incorporate smell micro-capsules in the polypropylene foam that release their contents when the helmet experiences a shock, such as would happen in an accident. The smell is sufficiently unpleasant to induce you to buy a new one; good for the cranium underneath, and good for the profits of the helmet makers. Smell-based fire alarms for the deaf make good sense. Vanilla-scented ice in the ice rink in the Eiffel Tower is said to be very popular, with Givaudan's *Vanille Givree* specially designed for the purpose. In several bus shelters in London, Nottingham, Glasgow, York, and Manchester in the UK you can press a button on advertisements for McCain's baked potato products, and receive a whiff of oven-fresh jacket potatoes, to warm you up as you wait for your bus. Adam's and Eve's noses may yet lead a marketing revolution.

*　　*　　*

How long will it be before law enforcement agencies keep a record of your individual scent signature on their files, much as they do with the fingerprints and DNA configurations of convicted criminals? It always used to be the case that the local policeman on the beat, who knew the local small-time villains from long experience, could enter premises and say, with certainty, that 'Jimmy the Hood' or 'Fingers Fred' was responsible, for they always left their smell. The technology exists today for crime scene officers to take a sample of air when they first start their work, and a group of Spanish researchers are developing a personal smell biometric technique, with 85% accuracy at present, to identify people by the smell of their hands.[149] As the technology develops we can expect more on this front in the years ahead.

---

[149] Rodríguez-Luján, I., Bailador, G., Sánchez-Ávila, C., Herrero, A., and Vidal de Miguel, G. 2013 Analysis of pattern recognition and dimensionality reduction techniques for odor biometrics. Knowledge-Based Systems 52:279–289.

What about the use of smells for controlling crowds, as Aldous Huxley predicted in *Brave New World*? Seats in football and other sports arenas could release calming scents, like Huxley's 'soma', the exact amount and configuration of which could be determined by the arena manager and released electronically from concealed vents under the seats. Could smells eventually replace tear gas for riot control in the streets? Could absenteeism in factories and offices, in which repetitive work is the norm, be reduced through aroma management? Lavender has been found to be effective in European leather factories, where the foetid smell of animal skins leads to workers taking much time off. What would the fragrance of jasmine or tuberose do to productivity in modern, high-stress offices? Could shops put mind-altering perfumes into their air conditioning systems, subconsciously enhancing a buyer's experience and subtly making them buy more product? It's already been shown that men are more likely to buy a men's magazine if they can smell the underarm ingredient androstenol at the time they are making their purchasing selections. We may soon discover a raft of scratch-and-sniff technologies attached to many common household consumables, carefully tailored to raise our state of arousal and making us more likely to buy.

Odour delivery technology has flowed through to the digital world and will soon arrive at a smartphone near you. The most ambitious use of this technology, at least in the non-military world, came in 2011 with the release of 'Sound Perfume Goggles' by researchers in Japan and Singapore, designed to 'boost your social life'. The goggles are heavy and old-fashioned with hemispheres the size of half ping-pong balls attached to the legs. Infrared detectors on the frames detect the presence of a similarly attired person and alerts your smartphone *via* Bluetooth. Your phone sends your name, number, and your choice of a scent selection, together with a selection of your favourite music, to the recipient's smartphone. Their phone connects with their glasses and your 'scent-card' is released from their half ping-pong balls, wafting around them and enhancing their memory of you. Their smartphone reciprocates and soon you're bathed in their 'scent-card' and music. Your phone automatically photographs them through a lens in the bridge of your goggles and you can recall their name and phone number, and scent, later by looking at your pictures. And all without a word being spoken! Would history have taken a different course had Cleopatra and Marc Antony been wearing 'Sound Perfume Goggles'? Who knows, but it's questionable if Shakespeare would have written a play about it.

Speculating even more expansively, advances in gene therapy technology may yet facilitate restoration of operability of the inactive VNO genes we have in our genome. If restored genes could be introduced into the olfactory membrane, or even genes from the olfactory genome of a dog, a whole new vista of opportunity would lie before the fragrance industry. We should be alert, however, to the possibility that such a technological development may create as many new social problems as it would offer sensory opportunities. And when genes can be designed in the laboratory and introduced into the genome, to express themselves in the olfactory membrane, the world of our descendants will be an utterly different one from that which we know today.

The possibilities for scent to be added to almost everything we buy are almost limitless. Micro-encapsulation of fragrances has seen to that. Clothes that change their smell as the day passes, releasing a different fragrance at different times of the day (a boon for the time-poor, upwardly mobile executive); breakfast cereal brand perfumes inspired by sun, wind, and ears of golden corn; floor coverings that emit fragrance with every footfall; cars fitted with devices that emit unpleasant odours when the time has come for a service, forcing you to turn them in so the foul smell can be extinguished; magazines lightly fragranced with delicate scents, or reeking of musks, depending upon their subject. All these things, and more, are just around the corner. They're a far cry from sea squirts sniffing the water around them, moths everting their coremata to attract belles from far away, or male musk deer buck calming a skittish doe at the start of the mating season. Or are they? I'll let you judge.

*   *   *

That tiny mutation, occurring 23 million years ago, set humans on an evolutionary course, the likes of which nature hadn't seen before.

It enabled our ancestors to live communally, and later in the vast conurbations of today, at high density and in relative harmony within each group, while preserving the genetic imperatives of the family. Our ancestors became carnivores, even though their teeth were quite unsuitable for cutting or tearing flesh. It wasn't only collaborative hunting that enabled them to gain the advantage offered by big game. They learned to cook. The cooking of meat combined nutritional with social evolution, resulting in our species being the only one that lives to eat, rather than eats to live. Going out to dinner and lingering over a fine wine, hardly compares to the unholy brawl that accompanies vultures tearing into a dead antelope, or lampreys gouging chunks from the body of a dead whale, even though the metabolic outcomes of both are much the same. Our species is unique in the animal kingdom in having made the vital biological function of feeding into a social activity, an art form garnished with complicated rituals and social niceties.

We have to thank the absence of the small plate of bone found in macrosmatic species, the *lamina transversa* that separates the expired from the inspired airflow, for allowing retro-nasal smelling to add so much to the pleasure of eating (see Fig. 4.10). What cooking did was to provide us with food richly laced with volatile odorants, just right for a sensory collaboration between taste and smell. Cooking has been around since the days when *Homo erectus* walked on Earth's grassy plains, between 1.9 and 1.8 million years ago. The Harvard anthropologist Richard Wrangham developed this timeline from study of refinements of *H. erectus'* skull, suggesting that the well-developed bony ridge along the crest of the skull was used by early pre-humans for jaw muscle attachment, but lacking in *H. erectus*, as cooked food needs far less chewing than raw food.[150] Not only was the bony ridge far less developed in *H. erectus*, but its teeth were smaller and finer than its ancestors', adding weight to the argument that *H. erectus* was the earliest of our ancestors to routinely cook its food. When food had to be cooked before being eaten, the newly minted social carnivore had taken the first step toward a uniquely

---

[150] Wrangham, R. 2009 *Catching Fire: How Cooking Made us Human*. Profile Books, London.

human social activity. From that time forward, it's just possible that everything good, and everything bad that has befallen humankind at its own hand, owes something to discussions over a shared meal.

*　　*　　*

At our journey's end, and empowered with what we know about the evolution of the human nose, are we any closer to answering the question 'what has made humankind'? Although scientists, theologians, and intellectuals have debated this question for thousands of years, there's no simple or single answer. Ultimately, of course, it's our brains that separate us from our nearest relatives — ours are three times the size of a chimpanzee's and, based on our body size and average primate brain size, six times larger than would be expected. Yet humans share more than 98% of their DNA with chimpanzees, so the gulf of difference between us and them is not down to any simple genetic analysis. Within our genome are regions designated as 'Human Accelerated Regions' (HAR), regions where mutations have occurred faster than elsewhere in the genome — a high mutation rate is an indication of active evolution in that set of genes. A group of genes that deal with motor control is included in one HAR, giving support to the view that manual dexterity and the use of tools contributed to what has made us human.

There's no question that our fully opposable thumbs, together with strong hands and dexterous fingers, have contributed greatly to our success as a species. It has been said that the first tools for slicing meat — flakes of flint — functioned like external teeth, opening up new food resources to our weak-jawed ancestors. Richard Wrangham argued that cooking made humankind, because cooking releases locked-up calories that can't be accessed if the meat is eaten raw. Other anthropologists see language, and a brain able to understand the minds of others, as the definer of humankind. Certainly, it is hard to imagine how our complex social relationships could be maintained without a sense of self-awareness, and a sense of how others might respond to our actions.

In the field of reproduction, humans are characterized by the presence of one or more children still dependent upon their parents when a new sibling is born — a situation that doesn't occur in chimpanzees. Calling in help from grandmothers and others past reproductive age is a uniquely human attribute, underpinning cooperative breeding as an adaptation that made us human. But sitting above all these is the notion that what has made humankind is the co-existent, inseparable union of biology and culture; evolution assisted, as it were, by something that's not encoded in the genes, and doesn't share the continuous, incremental change that occurs in attributes under the pressure of genetic selection.[151]

These ideas, and doubtless there are many more, contribute to what makes us human, but at a more fundamental biological level than all of them is the genetic mutation that enabled our distant ancestors to live gregariously in family groups, in a brand new lifestyle. Without 'ADAM' closing down the VN system, that would not have been possible, and none of the higher-order ideas about what makes us human would be relevant. Without seizing the advantages of cooperative hunting, made possible by 'ADAM', descendants of those distant ancestors of ours would still be living in forests, like so many of our zoological relatives. Breaking the link between smell and sex, which has supported life on Earth for over a billion years, was what started us on the road to humankind. Viewed against that cataclysmic adaptation, everything else is of secondary significance.

*   *   *

Considering the biological and cultural complexity of his fellows, the English poet Robert Herrick summed up humankind in just 28 words:

> 'Man is compos'd here of a twofold part:
> The first of nature, and the next of art:
> Art presupposes nature; nature she
> Prepares the way for man's docility.'[152]

---

[151] Calcagno, J.M., and Fuentes, A. 2012 What makes us human? Answers from evolutionary anthropology. Evolutionary Anthropology 21: 182–194.
[152] Robert Herrick, 'Upon Man'. From: 1891 Alfred Pollock (ed) *Works of Robert Herrick*. Lawrence and Bullen, London. Vol 1 p 191.

He saw humankind as a biocultural being, a blend of biology and art, with art reflecting biology leading to humankind's civil development. (It was only in the late 18th century that 'docility' came to be associated with submissiveness; until then its sense was the ability to be taught.) His 28 words are as much an epithet for the sense of smell as they are for the making of humankind, for onto the biologically evolved smelling equipment in the nose and brain has been superimposed a rich fragrance culture, subconsciously activating neural brain pathways that evolved when we were truly animals. It should be no wonder that allusions to smells are so well represented in literature and visual art because smell is the archetypical sense of emotion. Intellect is for the eyes and ears.

Although some would dispute it, humankind is the only species of animal to display emotion so readily, so frequently, and so fullsomely. In the rest of nature, animals need senses to provide information of importance to their survival. Is a grazing antelope 'frightened' when it spots a lion stalking it through the grass, as you and I would be? Or is its 'fright' response simply the implementation of a well-honed adrenalin system, kicking in to do its job at precisely the right time? If the system fails and the lion wins, the genetic aptitude to linger for another mouthful will not be transferred to the next generation. If the antelope wins, that genetic aptitude will survive a while longer. There is no need to invoke emotion — genes and Darwin's theory are enough.

Human lives are enriched by emotions; huge industries are built to pander to them — fragrance, film, and publishing industries, to name just three — and judging by the money we spend on perfumes, movies, and the print and social media, they are massively important to our well-being. It's through our noses, however, that we experience the deepest and most lasting emotional experiences. It's through our noses that we sense the world as it used to be, a complex mesh of interlocking and evocative scents. Smells of home that are never forgotten, wherever we may live; scents of dearly departed, perhaps hardly noticed during their lifetimes; aromas of flowers and forests, uplifting the most leaden of souls. Just a few molecules swept along on an incoming tide of air can trigger neural pathways in the brain once leading our ancestors to feed, fight, flee, or copulate, now take us to new levels of human fulfilment. The evolution of Adam's nose, and Eve's nose of course, is truly the story of the making of humankind.

Index